D0023083

THE SEVEN HILLS OF ROME

THE SEVEN HILLS OF ROME

A Geological Tour of the Eternal City

Grant Heiken, Renato Funiciello, and Donatella De Rita

PRINCETON UNIVERSITY PRESS

PRINCETON AND OXFORD

Published by Princeton University Press, 41 William Street,
Princeton, New Jersey 08540
In the United Kingdom: Princeton University Press,
3 Market Place, Woodstock, Oxfordshire OX20 1SY

Library of Congress Cataloging-in-Publication Data

Heiken, Grant.
The seven hills of Rome : A geological tour of the Eternal City / Grant Heiken,
Renato Funiciello, and Donatella De Rita.
p. cm.
ISBN 0-691-06995-6 (alk. paper)
1. Rome (Italy)—Guidebooks. 2. Geology—Italy—Rome—Guidebooks. I. Funiciello, R.
II. De Rita, Donatella. III. Title.
DG804.H445 2005
913.7′602—dc22 2004016569

British Library Cataloging-in-Publication Data is available

This book has been composed in Minion Text

Printed on acid-free paper. ∞

pup.princeton.edu

Printed in the United States of America

10 9 8 7 6 5 4 3 2

Contents

Foreword, by Walter Veltroni vii

Preface ix

Chapter 1
A Tourist's Introduction to the Geology of Rome 1

Timelines 18

Chapter 2
Center of the Western World—The Capitoline (Campidoglio) Hill 27

Chapter 3
Palaces and Gardens—The Palatine (Palatino) Hill 37

Chapter 4
The Aventine (Aventino) Hill 51

Chapter 5
The Tiber Floodplain, Commerce, and Tragedy 59

Chapter 6
The Tiber's Tributaries in Rome—Clogged with Humankind's Debris 85

Chapter 7
The Western Heights—Janiculum, Vatican, and Monte Mario 110

Chapter 8
The Celian (Celio) Hill 123

Chapter 9
Largest of the Seven Hills—The Esquiline (Esquilino) 153

CHAPTER 10
Upper Class—The Viminal (Viminale) and Quirinal (Quirinale) Hills 162

CHAPTER 11
Field Trips in and around Rome 174

The Seven Hills of Rome in Fifteen Stops 174
Panoramas, Piazzas, and Plateaus 195
A Field Trip to Rome, the City of Water 216

Acknowledgments 229

Further Reading 231

Index 237

Foreword

FROM ITS TIME as the historic center of the Roman world, Rome has been continuously a political, religious, and administrative capital. Geologic and terrain factors have assured its population growth and, above all, provided the conditions for survival of the most modern culture in the ancient world. From lessons of urban development and prosperity, the Roman people developed the capacity to recognize and to manage in a positive way the natural resources of the region. The volcanic terrain, the Tiber River and its complex watershed, the water resources of the central Apennines and surrounding countryside, and the abundant natural materials for construction, roads, and aqueducts have all contributed to the birth, growth, and success of Rome. If these natural riches were to be considered *res nullius* (a thing that has no owner), the city would experience progressive and eventually fatal decline. Worsening political and economic fortunes would be accompanied by inattention to the wonderful but fragile equilibrium of natural factors and their importance for urban development. Calamitous events, including the depletion of natural resources, would all contribute to a scenario of decline more profound than that of the ancient city in the 5th century A.D.

Modern Rome is a new city, born after the unification of Italy and developed in an impetuous manner because of the complex historical backdrop of Italy and Europe. Many aspects of the modern urban environment are leading to dissatisfaction and suffering for many, stimulating a strong awareness of the need to care for the city with great attention to its management and with appreciation of its environment. The long history of our city and links to its setting and environment are of increased interest to the international community. It is appropriate that the term *urban geology* has its origin in *Urbs*, which was the ancient name for the city of Rome. Justifiable interest in urban science is growing in the international community of scientists and in city administrations. City residents must understand the roots and factors both hidden and manifested by the environment in which they live.

The authors of this work have offered the citizens of Rome and its visitors a new understanding of the city and its environment, based on a synthesis of work by scholars and city officials. I hope that *The Seven Hills of Rome* will stimulate other cities to examine their environments in the same way.

WALTER VELTRONI
Mayor of Rome

Preface

A sunrise over Rome, as seen from the Monteverde area above the right (west) bank of the Tiber, is a truly memorable experience, with the mixed clouds, a bit of mist above the river, and bells ringing everywhere. The distant Apennines and Alban Hills form a backdrop for center stage—the seven hills of Rome. Monteverde and the nearby Janiculum are wonderful viewpoints from which one can contemplate the history and cultural significance of this great city as well as its geologic foundation. Across the river, the seven hills are actually the eroded remnants of a plateau dissected over many centuries by streams. The stream valleys are less evident because man-made debris from more than three millennia has smoothed the original terrain.

Why was this site chosen for Rome of the Republic, Rome of Imperial times, and eventually Rome of the popes and Rome, the capital of unified Italy? For many reasons: proximity to a major river with access to the sea, plateaus affording protection, nearby sources of building materials, but, most significantly, clean drinking water from springs in the Apennines. Most of us recognize Rome as a source of inspiration for historians, architects, artists, musicians, and theologians. However, it's interesting to note that the success of Rome would not have been possible without the myriad benefits provided by its geologic setting. Even the resiliency of its architecture and the stability of life on its hills are supported by its geologic framework.

In addition to supplying an ideal setting, geologic processes have also threatened Roman life and property with floods, earthquakes, landslides, and (in the Bronze Age) volcanic eruptions. Understanding these threats allows planners to best situate homes and civic buildings and to lower the risk from the natural events so common on our dynamic planet. Rome is uniquely fortunate in that it has an excellent historical record that reaches back to Rome of the Republic—a record from the

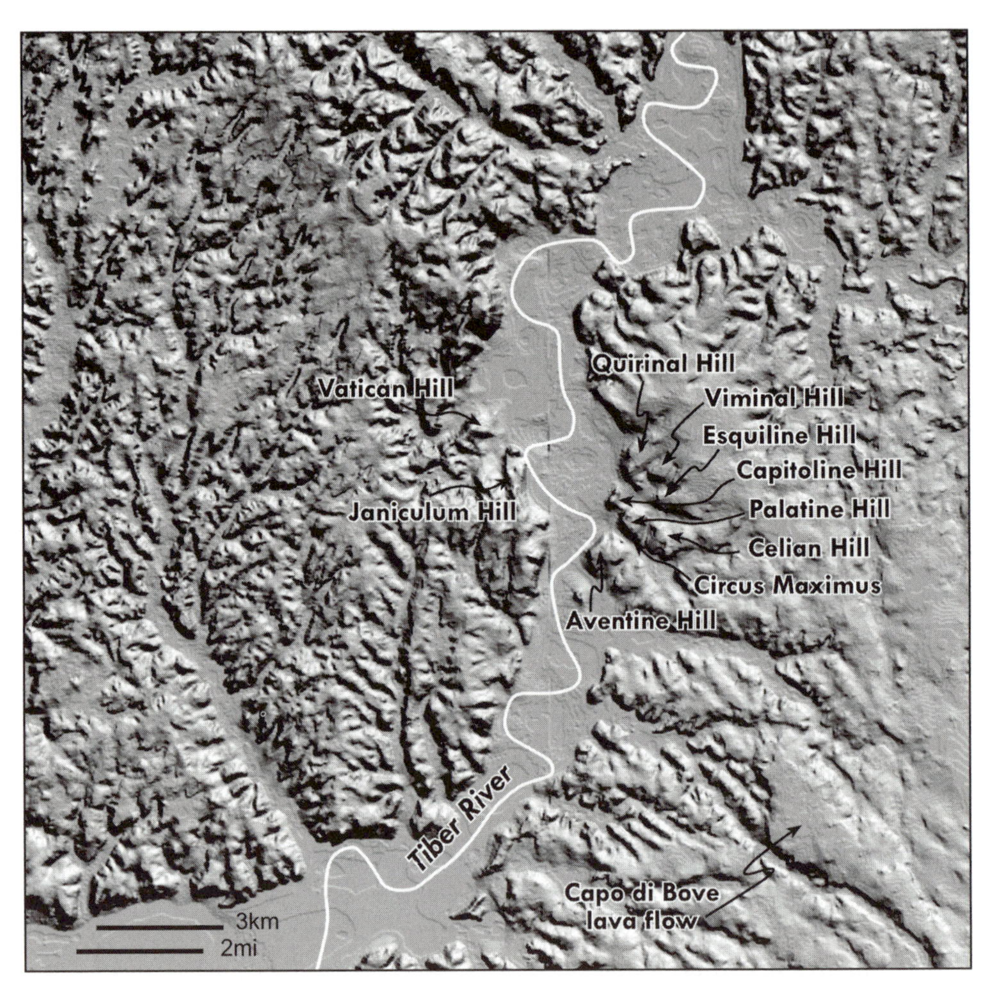

The terrain with its underlying geologic framework has played an important role in the success of Rome as a city and an empire. The Tiber River, crucial to water transport into ancient Rome, was trapped on the west by the fault blocks of the Janiculum and Vatican hills. The Tiber's east banks are volcanic plateaus, composed of deposits from explosive eruptions in the Alban Hills. In addition to forcing the Tiber toward the west, these deposits make up the "seven hills of Rome."

past that, in turn, has helped geologists and geophysicists in their efforts to mitigate future disasters.

This journey began in 1995 with the publication of a technical monograph, *Geologia di Roma* (*The Geology of Rome*), by the National Geological Service of Italy. The work represented in the monograph should serve as a standard to be followed by every large city in the world. Understanding a city's geologic underpinnings and processes can save lives and property, guide development, and secure a sustainable future. The scientific coordinator of the monograph, Renato Funiciello, was also one of the thirty-seven contributors. At about that time, Grant Heiken was involved with the Urban Security Project at the Los Alamos National Laboratory, the purpose of which was to understand the interconnectivities of systems that make up a city. Funiciello and Donatella De Rita gave Heiken a copy of the Rome monograph at the "Volcanoes in Town" meeting in September 1995. The monograph, the meeting, and long discussions motivated the three of us to prepare a book, for the public, on the relationship between geology and the history of Rome, to demonstrate the importance of geologic setting in a city's past and future.

The book may be read by itself or as a supplement to histories of Rome or carried, along with a good city map, to expand a reader's understanding of the city as he or she explores its streets. Geologists love field trips, and we hope to infect the reader with an equal enthusiasm through the guides in the last chapter.

The book's chapters are arranged geographically and cover each of the seven hills, the Tiber floodplain, ancient creeks that dissected the plateau, and ridges that rise above the right bank. The chapters may be read sequentially or randomly.

Chapter 1 provides an overview of Rome, using views of the geologic setting at scales of a few tens of meters to thousands of kilometers and at timescales of hundreds of millions of years to a few hundred years. A basic description of the geologic history of the Mediterranean is included.

A visit to the Capitoline Hill in chapter 2 focuses on the eruption products of the two major volcanic fields that flank the city. Chapter 3 describes the Palatine Hill and introduces the reader to the consolidated volcanic ash deposits (tuffs) that form the seven hills and have provided most of Rome's building materials. The peaceful Aventine Hill is the venue in chapter 4 for discussing the city's defensive walls, as well as the hazards and opportunities provided by the quarries and catacombs that underlie the seven hills.

The importance of the Tiber to Rome and its commerce, in addition to the threat of floods it brings, is covered in chapter 5. This section also highlights the discovery of concrete by Roman engineers and the subsequent importance of that material to the city and the empire. Chapter 6 explains how the debris of millennia that underlies Rome is a bit of a curse to a modern city but a treasure trove for archeologists. Chapter 6 also covers the risk of earthquakes to the Eternal City and how ancient monuments help reveal damage caused by past earthquakes.

Ancient Roman roads enter the city from all directions; in chapter 7, the reader is introduced to the hills west of the Tiber (the Janiculum, the Vatican, and Monte Mario) and then approaches the city by the Via Aurelia, which crosses these ridges along the right bank of the Tiber. This chapter discusses both the risk of landslides and the rather mundane but critical resources of sand, gravel, and clay that were needed to build this great city.

Rome, one of the few cities in the world that has an excellent supply of drinkable water, has flourished on the basis of available water: history reveals that when the aqueducts were damaged, Rome's fortunes declined. A visit to the quiet Celian Hill in chapter 8 introduces the reader to water resources and to the travertine (formed by springwater deposition) commonly used in construction and sculptures. Chapter 9 takes the reader across the Esquiline Hill, one of the largest of the seven hills, where the topic is energy in ancient and modern Rome—the sources and use—and lessons we can learn from the past.

The importance of their geologic foundations in the survival of the commemorative columns of Trajan and Marcus Aurelius is described in chapter 10. The marble so prevalent in the architecture and artworks for which Rome is renowned, comes from many sources around the Mediterranean. The final chapter consists of three field trips around the center of Rome, which can be taken on foot or by public transportation. Any reader who completes all three trips deserves a good Roman meal and a bottle of wine!

We hope that after reading about the remarkable link between the geology and history of Rome, readers will take a hard look at their own urban communities. An understanding of every city's geologic framework is not only interesting but also important in planning for the future of that city.

THE SEVEN HILLS OF ROME

The Trevi Fountain was built on the north end of the Trevi Plaza between 1732 and 1762 by a large team led by Nicola Salvi. Attached to the facade of the Palazzo Poli, it is one of the most extravagant of Rome's many fountains (and one of the largest, at more than 30 meters wide). Water was transported into the city's public plazas through aqueducts originally built by ancient Romans and restored after medieval times. The water that flows so freely in the Trevi Fountain comes from springs near Salone outside Rome, via the Vergine aqueduct.

CHAPTER 1

A Tourist's Introduction to the Geology of Rome

At early midnight, the piazza was a solitude;
and it was a delight to behold this
untamable water, sporting by itself
in the moonshine.
—Nathaniel Hawthorne, *The Marble Faun*

The monumental Trevi Fountain in central Rome symbolizes the relationship between the city and its geologic underpinnings. The stone from which sculptors created this work of art, the clean water from springs in the Apennines and volcanic fields near the city—transported by the famous Roman aqueducts—and the stones underfoot are all products of Rome's geologic heritage.

Construction of the fountain began in 1732, following a design by Nicola Salvi and using stone from the region. Travertine, a sedimentary spring deposit from quarries near Tivoli, and marble, a metamorphic rock from Carrara, in northern Italy, were used for the figures. The plaza is paved with small blocks of lava from flows along the Appian Way. For more than two millennia, Rome's fountains have provided neighborhoods with clear, refreshing water from springs in the Apennines, the Alban Hills, and the Sabatini region: a precious resource transported through aqueducts that were built during the Roman era and restored by the popes beginning in the 16th century.

This is your first visit to the Eternal City of Rome and, with guidebook and map, you plunge into its historic center. The goal is the Trevi Fountain, one of Italy's most famous landmarks. The trek can be daunting. Myriad small piazzas are connected with narrow streets, twisting this way and that, cars and scooters crowd the pavement, and the modern Roman phalanx—a tour group—impedes your progress. Buildings of all shapes and vintages block your horizon, scaffolding masks the architectural lines of famous landmarks, and resurfacing hides the ancient streets, making it impossible to view the city's past, hidden under its many debris layers.

During a brief visit, how do you get a grip on the geographic and temporal components of Rome, where a remarkable combination of geologic setting, environment, and history has produced a city that attracts millions of visitors every year? One fascinating approach is to imagine that you are able to rise above the Trevi Fountain, pausing at different elevations above the city so you can see Rome through a series of windows: first, just 30 meters, then 300 meters, then 3, 30, 300, and 3,000 kilometers on a side. Examining the setting of Rome from these six perspectives allows us to view the interactions between geologic setting, urban development, natural disasters, and humans' continuing struggle to modify and control the environment

The 30-Meter Window

Approximately 30 meters (98 feet) wide, the Trevi Fountain dominates its small piazza and is one of Rome's most easily recognized landmarks. Most movies filmed in Rome include the requisite scene at (or in) the fountain. Tour leaders and books remark on its ornate sculptures and the way that both the figures and the water emerge from the rock. The piazza actually is a small area, but even at this 30-meter scale, we can learn quite a bit about the importance of geologic setting in the history of Rome and its Empire.

To begin with, why is such a large fountain located in such a claustrophobic space? Seeing it for the first time, visitors are frequently amazed that such an astounding monument is seemingly tucked into a corner of a crowded city. It's important to remember that, despite their sometime glorious appearances, Roman fountains for 2,400 years served the practical purpose of providing water for the populace. A neighborhood fountain supplied this precious fluid for drinking, cooking, cleaning, and flushing public toilets. During the Republican period and the Imperial dynasties, Rome had an abundant supply of clean water from several sources, thanks to its geologic setting and extraordinary engineering. The water infrastructure was later rebuilt and restored under the popes.

The Trevi Fountain occupies most of this small plaza. The Trevi was as much a display of art as a source of water for the neighborhood, and its light color reflects the use of travertine and marble in its construction. Although not easily seen here, the streets and plaza are paved with *sanpietrini*, small blocks of lava quarried from lava flows in the Alban Hills, a volcanic field southwest of Rome.

The Trevi Fountain, among others, was and still is supplied by the Vergine aqueduct (Aqua Virgo), which brought water from springs at Salone, 16 kilometers (10 miles) east of central Rome, via a circuitous route that enters the city from the north. Inaugurated in 19 B.C., the aqueduct was damaged during the siege of Rome by the Ostrogoths in A.D. 537–38 and was reconstructed near the end of the 15th century. Most of the Vergine aqueduct is underground and passes immediately under the Piazza Trevi. Three streets converge at this fountain, so it is possible that its name may have derived from *tre vie* (three streets).

The first fountain at Trevi was a utilitarian model, built for Pope Nicholas V in 1453 and derided as the "village well." Bernini had this fountain destroyed in anticipation of erecting one of his own design. In fact, his design was not used, but his influence resulted in the fountain being moved from the south to the north side of the piazza, its present location. After an intense competition between sculptors in 1730, the design of Nicola Salvi was selected. Construction of the new fountain took thirty years, between 1732 and 1762, using two architects, ten sculptors, and many assistants. The fountain's travertine base emulates nature, with rough stones, cascades, crevices, grottoes, and carved representations of thirty plant species. The figures, including Oceanus (Neptune) and the Tritons, are carved in Carrara marble, one of the finest natural materials used by the greatest sculptors.

Although the fountain once supplied fresh water to the neighborhood, the flowing cascades are now recirculated and are no longer potable. If you're thirsty, however, *fontanelle* (small water fountains) along the shallow steps leading down to the fountain provide clear, cool, drinkable water.

Tired? The Trevi's steps are an excellent place to sit for a while and look around. The rounded paving stones below your feet are *sanpietrini*, blocks cut from lava that flowed from one of the volcanoes of the Alban Hills to the edge of what are now Rome's city limits. These stones are the same type Imperial Rome laid down for heavily traveled roads throughout its empire.

Although you can't see it, beneath the *sanpietrini* there is plentiful evidence of both anthropogenic (human-related) and geologic events. Immediately below is a 5- to 10-meter-thick (16- to 30-foot) layer of debris left by man's activities; it is mostly within these debris layers that archeologists find clues to the city's complex history. Below the debris is a 60-meter-deep (197-foot) channel cut by the Tiber River as it flowed into a sea much lower than today's Tyrrhenian Sea. Sea level has since risen during the latest warm cycle of the Earth's atmosphere, and the Tiber valley has been subsequently filled in with river sands, gravel, and mud. Beneath this alluvium is a thick sequence of fossiliferous sandstone and mudstone layers that were deposited in an ancient seabed 2 to 3 million years ago.

We could go still deeper, but we'll stop here, let you catch your breath, then return to street level and begin our rise above the Trevi Fountain's neighborhood.

More about the Stone Used in the Trevi Fountain

Peter Rockwell, an American sculptor living in Rome, is an expert on the history of stone carving. When he analyzed the features, sculpting techniques, and construction of the Trevi Fountain, he found that the fountain is 89.2 percent travertine, 7.2 percent marble, and 3.6 percent travertine breccia. The principal stone used for the base of the fountain (the scogliera da sola) is travertine, a porous calcium carbonate spring deposit. Roman travertine was (and continues to be) quarried near Tivoli, east of Rome, where bicarbonate mineral warm water issues from springs found along faults at the base of the Apennines and flows into a sedimentary basin.

Travertine is a particularly useful rock type: for the geologist, it provides clues to the dynamic history of the Apennines and adjacent sedimentary basins; for the hydrologist, it reveals information about the evolution of the springwaters; and for the archeologist or art historian, it contributes to the provenance of many sculptural pieces.

Travertine breccia was at one time a uniform, thin-layered, brittle spring deposit that was broken by faults. The angular pieces of rock were then cemented by younger travertine as water flowed through the rubble—the final product is a "breccia." The famous Carrara marble is a metamorphic rock (limestone that has been altered by high temperatures and pressures) from northwestern Italy.

The 300-Meter Window

Centered on the Trevi Fountain, a 300-meter-square (984-foot) window offers a view that includes parts of the Trevi and Colonna neighborhoods. Immediately east of the fountain, the natural terrain rises 40 meters (134 feet) until it meets the lower walls of the Quirinal Palace. The Quirinal Hill, one of those famous "seven hills of Rome," was a residential area in Imperial Roman times, was the site of the pope's

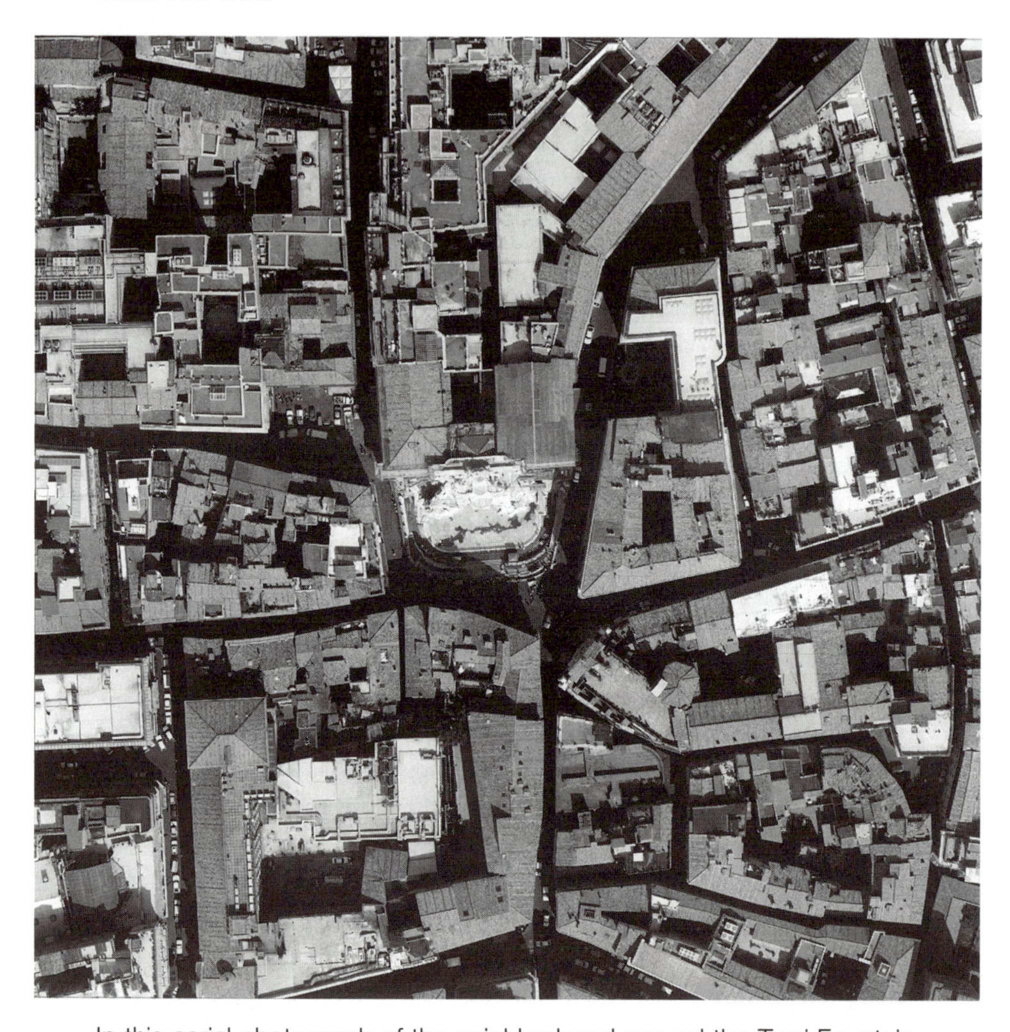

In this aerial photograph of the neighborhood around the Trevi Fountain, the edges of the image are 230 meters (750 feet) by 260 meters (850 feet). To the west (left) of the fountain, the north-south streets overlie sediments of the Tiber River. The curving street to the east may follow a drainage at the base of the Esquiline Hill, located at the right edge of this photo, which consists of deposits of volcanic rock (tuff). All the original geologic features have been masked by accumulations of debris over the millennia.

summer palace, then the home of the Italian royal family, and, most recently, the official residence of Italy's president.

Much of this area is underlain by the sands and muds of an alluvial plain deposited when the Tiber overflowed its banks. Until the 1950s, the Tiber regularly ravaged central Rome with floodwaters that reached as far as the lower Via del Tritone—just beyond the northwestern edge of this view.

The Quirinal Palace was constructed on the edge of a plateau; the flat area was built up from the alluvium and marsh deposits of an early Tiber River, which in turn were overlain by deposits of consolidated volcanic ash from the Alban Hills and Sabatini volcanic fields. These volcanic ash deposits were deposited by fast-moving flows of hot gas and ash from eruptions between 600,000 and 300,000 years ago. Blocks from consolidated ash deposits (tuffs) have been used throughout the history of Rome (and, indeed, throughout the world) as a common building stone. There is ample proof that tuff deposits also offer a stable foundation for construction; overlying buildings have been minimally affected by Rome's earthquakes.

In this view we can see that the Tiber's tributaries have cut ravines and small valleys through the Quirinal Hill. These erosion channels, as well as the Tiber's ancient channel, were most likely carved when sea level was lower and are now partly filled with alluvium. The Via del Tritone, mentioned previously, follows what was an alluvium-filled ravine that has also been partly filled in with man-made debris.

Our geologic information about Roman sites is based on extremely rare outcrops, underground quarries, and engineering drill holes. Geologic mapping within a city is always a challenge because so much terrain is covered with the debris from several millennia of human activities. Fortunately for us, many Roman and Italian organizations have spent decades producing an interdisciplinary study of the geology of Rome.

The 3-Kilometer Window

Pulling back farther gives us a 3-kilometer-wide (1.86-mile) window through which we can view a large piece of Rome's historic center, including all of the famous "seven hills." To thoroughly explore this 9-

square-kilometer (3.5-square-mile), densely packed city center with its varied, complex history in a single week would challenge even the most dedicated tourist. From this vantage point, however, the geologic framework of the city becomes more understandable.

The tuff plateau, with its seven hills, is easy to identify on a relief map; it consists of a sequence of ancient sedimentary rocks left by the Tiber and volcanic rocks (tuffs) from the Alban Hills and Sabatini volcanic fields. Here are the Quirinal, Viminal, Esquiline, Capitoline, Celian, Aventine, and Palatine hills, as well as the Pincian, which is part of the same plateau that now hosts the vast Villa Borghese Park. Many ancient Roman ruins occupy these hills, the most famous of which are visible near the Roman fora and the Colosseum. Massive tuff deposits from volcanoes changed the course of the Tiber and narrowed its valley floor to create what became a strategically located city that could be strongly defended but had easy access to water transportation. The floors of small tributaries were convenient, open sites for markets, theaters such as the Theater of Marcellus, and larger public structures like the Colosseum and the Pantheon. The plateau's tuff deposits were also the source of stone used for early city walls and the foundations of the great buildings of Imperial Rome.

Development of a densely packed city on the Tiber floodplain began during medieval times. The plain, once occupied chiefly by Roman theaters, temples, and army training facilities—all easily cleaned (and repaired) after a flood—now began to accumulate homes and businesses as well. Floods submerged such built-up areas as the now-well-known Piazza Navona and the Trastevere neighborhoods. If these later generations had followed the urban planning strategies of their ancestors, there would have been far less damage and loss of life during post-Imperial city growth. Planning ways to mitigate the effects of flooding was a standard process for early Romans—one that should be adopted even today in the world's cities.

The 30-Kilometer Window

Looking down at Rome through a window 30 kilometers (18.6 miles) square, we can see most of the modern city, its suburbs, and the ring road (Gran Raccordo Annulare). This view extends well beyond the

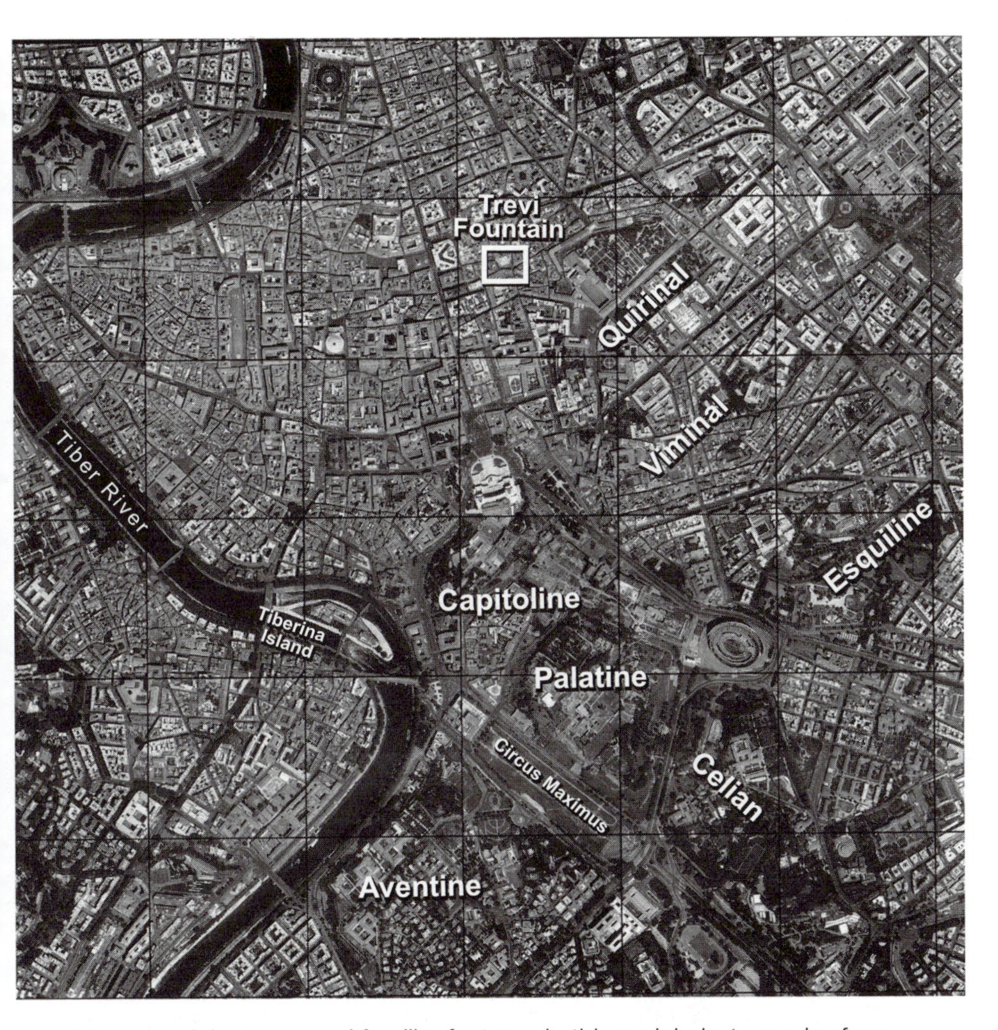

You can pick out several familiar features in this aerial photograph of central Rome (3 kilometers on a side), which is centered on the Trevi Fountain (white rectangle). Most of the more important elements of Rome's geology also can be seen here. The "seven hills" are visible, including the Quirinal, Viminal, Aventine, Esquiline, Celian, Palatine, and the Capitoline (the last four of these surround the Roman fora and the Colosseum). All the hills are erosional segments of a plateau that consisted of mostly volcanic tuffs that were erupted in the Alban Hills to the southeast of Rome. The flat floodplain of the Tiber and several of its meanders are visible on the left, as is the elongate Tiber Island. Tributaries of the Tiber drained the plateau and left ravines and small valleys that are now partly filled with debris.

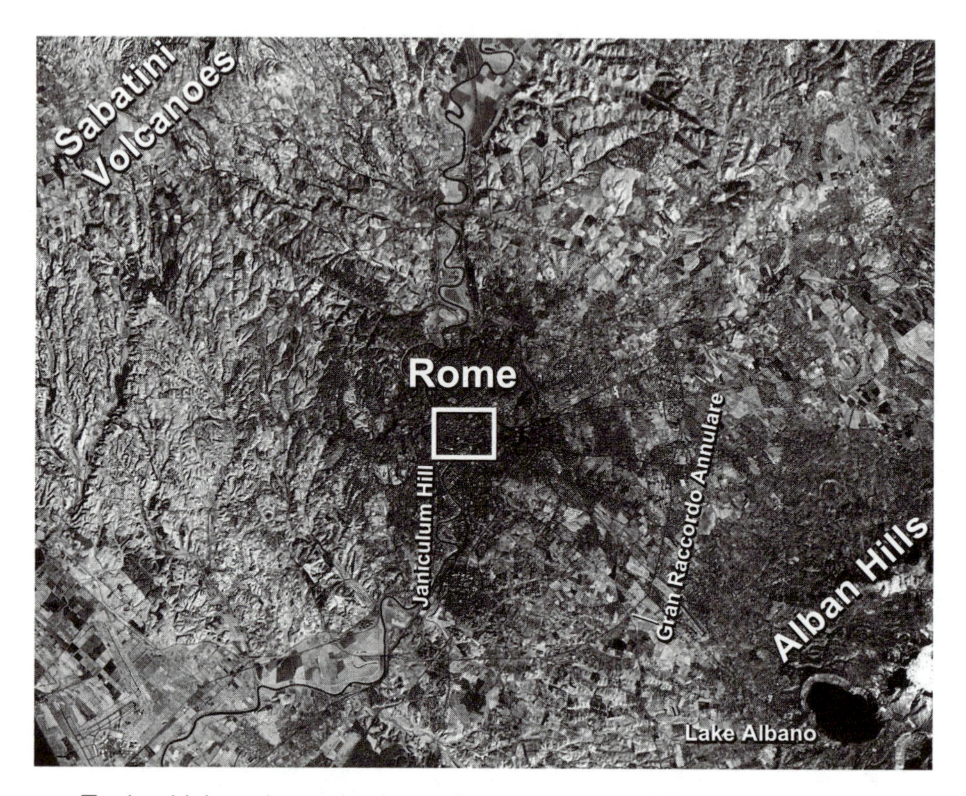

To rise higher above the immediate area around Rome, we needed a satellite image; this one is approximately 30 kilometers on a side. Much of 20th-century Rome is included within the Gran Raccordo Annulare (GRA), the faintly visible ring road that circumnavigates the city and passes over many of the geologic components of Rome. A north-south-trending block of 2 million- to 700,000-year-old marine sedimentary rocks defines the eastern edge of the Tiber floodplain as it passes through Rome. From the southeast and northwest, volcanic plateaus slope toward the Tiber from their sources in the Alban Hills and Sabatini volcanic fields. The floodplain of the Tiber is narrow here, having been confined by a north-south-trending basin and volcanic rocks before it turns west below the Janiculum and flows into the Tyrrhenian Sea. The area in this image encompasses much of what was Rome before the numerous conquests that created the Empire.

original walls of the Imperial city, nearly reaching an area under control of the Etruscans in what was the city of Veii, 16 kilometers (10 miles) north of Rome.

The Tiber River follows a structural depression created late in the geologic history of the region, when the land was being pulled apart by movements of the Earth's crust. At this scale, we see the Tiber crossing Rome from the north, then turning southwest toward the Tyrrhenian Sea. The hills west of the river (Monte Mario and the Janiculum [Gianicolo]) are composed of million-year-old marine mudstones and sandstones, evidence of a time when the region was beneath the sea.

Eruptions in volcanic fields located southeast and northwest of Rome created two plateaus that descend toward the Tiber. Rapidly moving flows of ash and gas from explosive volcanic eruptions dammed the river with deposits of ash (tuffs) and changed its course. Both of the volcanic fields, the Sabatini to the northwest and the Alban Hills southeast of Rome, have played an important role in creating the terrain that we see today: gentle plateaus pinching the Tiber floodplain and creating high ground for the city. In geologic terms, the volcanic fields are young—the most recent eruptions occurred in the Alban Hills about 3,500 years ago. Roman writers such as Livy and Pliny the Younger recount tales of explosions and "rains of fire" in the Alban Hills, but none of these events have been verified. A recent study by two of the coauthors supports the hypothesis that large bursts of gas in Lake Albano in historic times buried sections of the volcanic slopes with mud.

As the Tiber River leaves central Rome and the narrow valley that was created between upfaulted sedimentary rocks and volcanic ash plateaus, the slope of the riverbed decreases and the flow is placid as the river approaches the sea. This is the head of the Tiber delta, which has been expanding into the sea since the river first reached the coastline. The major Imperial Roman seaport of Ostia (now Ostia Antica), near the window's southwest corner, is now 4 kilometers (2.5 miles) from the sea, landlocked by the silts and sands of the growing river delta. The Tiber and its delta were key factors in the mercantile and military successes of the Roman Empire, making it possible to establish ports near Rome and thus ship materials and goods upriver into the city.

The 300-Kilometer Window

At the 300-kilometer (186-mile) scale, which encompasses all of the Italian Peninsula, we can see many geologic components that have affected the history of Rome and its interactions with other peoples. Terrain had a great impact on the early growth of the Roman Empire and still influences Italy's natural resources and transportation systems. We'll look at several of these features more closely.

The Apennines

The Apennines, the backbone of the Italian Peninsula, have had both historic and geologic importance. These rugged limestone peaks and ridges run from Genoa and Turin down to the instep of Italy's "boot" south of Taranto. Above the timberline, the rugged snowcapped peaks provide rocky platforms for everything from ski areas to radio and television transmission antennas. The highest of these peaks is the Gran Sasso, with an elevation of 2,912 meters (9,554 feet). Because the foothills of the Apennines are so close to sea level, the chain is truly impressive as it rises abruptly above the Tyrrhenian coastal plain to dominate Rome's eastern and northern horizons. High mountain valleys host small farms and many villages. Today only a few large highways are visible in aerial photographs; many of the main routes follow deep tunnels cut through the mountains.

The ridges of the high Apennines, with their harsh alpine terrain, separated the cultures and languages of pre-Roman Italy. The high ground was held by the Sabines, Etruscans, Lucanians, and Peucetians; the Adriatic coastal plain by the Daunians and Picenes; and the Campanian Plain by Campanians and Greeks. Strongly independent, these peoples were isolated by terrain and language. To defend itself and expand its empire, Rome had to conquer the rugged natural barriers and then establish colonies of military veterans whose presence helped pacify the indigenous population of those distant valleys.

Roman consular roads connected the colonies by following creases in the natural terrain that were created by folds and thrust faults (slices of rock pushed over one another to create what looks like a slanting

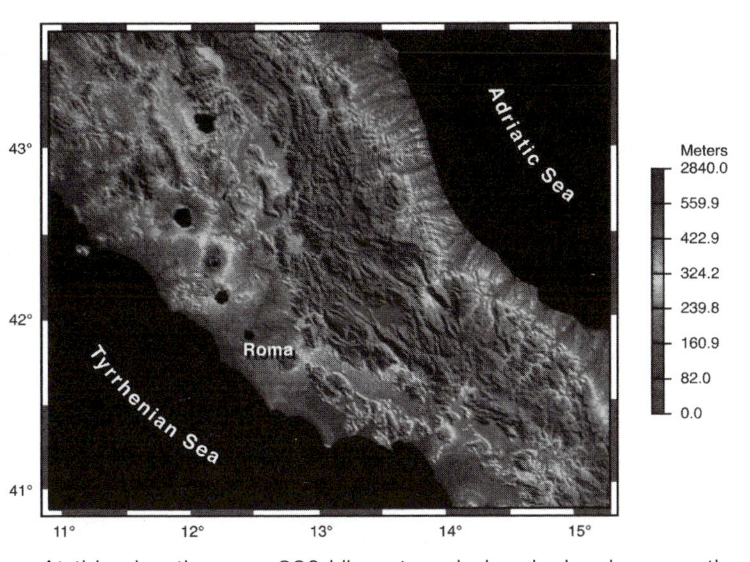

At this elevation, our 300-kilometer window looks down on the entire central Italian Peninsula, from the Adriatic to the Tyrrhenian Sea. The northwest-southeast-trending Apennine Mountains, which form the backbone of the Italian Peninsula, are easily seen in this digital elevation map. The high mountain chain is the product of tens of millions of years of geologic compression and the stacking up of slices of the Earth's crust. In contrast, lowlands on the western side of the Apennines were formed during extension, or pulling apart, of the crust over the last several million years. The oval or circular mountains near Rome (with large crater lakes) are volcanic fields.

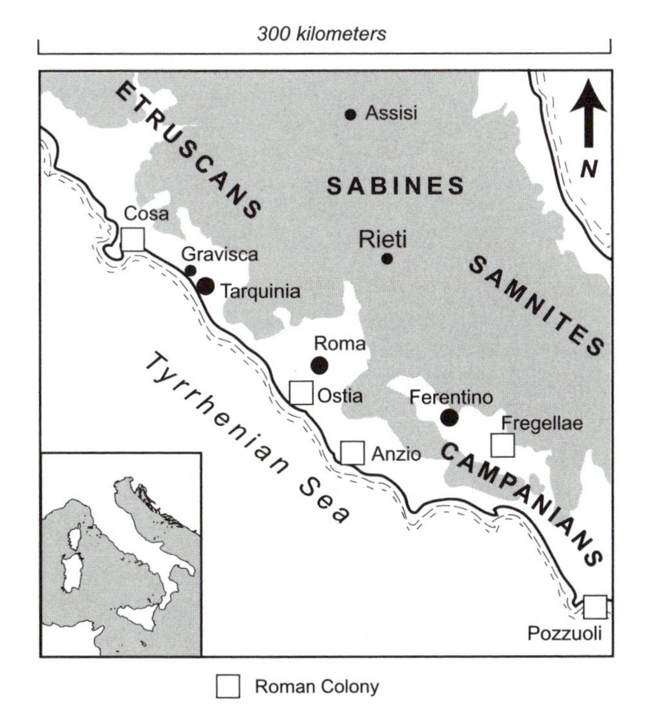

The rugged, high peaks of the Apennines were occupied by many different cultures with multiple languages. Both the mountains and their peoples had to be controlled before Rome could develop its empire. (Adapted from Potter 1987)

deck of cards). For example, both the Via Flaminia and the Via Appia follow routes along valleys formed by erosion along major faults. When following the structural grain was impossible, well-built mountain roads traversed passes between valleys.

Limestones deposited in ancient shallow seas as long ago as 300 million years have become a very important component in the Eternal City's sustainability: they formed the major rock reservoirs that hold Rome's water supply. The limestones are also quarried for building stone and provide the lime used in concrete.

You might ask, "How could all these marine sediments end up being exposed in a chain of high mountains?" The Apennines are the product of a collision between the continental plates of Africa and Eurasia. Instead of forming a continuous, more or less east-west boundary between the two continents, as one might expect, however, the merging developed on a trend from the northwest to the southeast, along the shoreline of the present-day Adriatic Sea and the eastern slopes of the Apennines.

The complex collisions and extension (pulling apart) of the Earth's crust in this region lasted from about 20 million to 2 million years ago. The compression (or "thrusting") occurred along two major lines that run along the Italian Peninsula; these are separated by a north-south-trending fault, along which the motion is absorbed by strike-slip movement (two blocks moving laterally along a nearly vertical fault). At the same time, the Italian Peninsula began to rotate counterclockwise, opening basins to form what is now the Tyrrhenian Sea.

The Tyrrhenian Sea and Coastal Plain

These cycles of extension and compression not only thinned (or stretched) the oceanic crust below the Tyrrhenian Sea but also produced high heat flow and volcanic activity beneath the sea and along the Tyrrhenian shore. The coastal plain we see today follows the transition between the thrust faults of the Apennine chain and the subsiding Tyrrhenian basin, which has been opening slowly over the last 20 million years. Formation of this coastal margin was irregular, beginning in the south and migrating toward the north.

Today, these complex processes of compression and extension continue to produce irregular movement along faults of all types on the Italian Peninsula, generating frequent earthquakes, some of them catastrophic. A further and even more dramatic activity—volcanism—has contributed to the Roman area's complex geologic history.

The Volcanoes of Rome

Volcanism is a natural phenomenon not often associated with Rome itself, but it has played a very important role in the formation of the Tyrrhenian coastal plain. From this viewpoint, we can see a line of volcanic fields along the edge of the Italian Peninsula from Naples through Rome and on to Tuscany. Much of this volcanism was explosive, producing plateaus that radiate from large craters, which are visible on this image as a number of circular or semicircular basins scattered from the Alban Hills to Bolsena Lake. Many of the large craters contain lakes that are reservoirs for some of the Roman water supply. The largest around Rome is Bracciano, which provides potable water and in the future may generate electricity from geothermal systems.

These are young volcanic systems (the last eruption was about 3,500 years ago). The largest of the eruptions left collapse craters (or *calderas*), which were formed when eruptions evacuated underlying magma chambers. Looking closely, we can see that superimposed on the calderas are smaller volcanic craters and some lava domes.

The Tiber and the Paleo-Tiber

If you approach Rome from the north, the major river you see is the Tiber. Upstream from the city, the river has a fairly wide floodplain and empties a drainage area that extends into north-central Italy. The Tiber follows the geologic structure that we discussed earlier and does not turn toward the sea until it reaches Rome. Before volcanic eruptions built up the fields flanking Rome, the ancient Tiber meandered along a structural basin, the western margin of which is formed by the hills behind the Vatican and Trastevere. The ancient river flowed to the south-southeast, below what is now Cinecittà—large, blocky buildings that house film studios at the southeastern edge of the city. Today's

Tiber flows through Rome along a narrow valley; its path was shifted to the west by rapid deposition of volcanic ash from eruptions in the Alban Hills.

The 3,000-Kilometer Window

At 3,000 kilometers (1,863 miles), our panorama frames most of the Mediterranean—one of the most geologically complex areas of the Earth. Now we see the consequence of movements between the African and Eurasian plates of the Earth's mobile crust. Slowly, slowly, the African plate is sliding under the Eurasian plate. Geologic events created today's Mediterranean Sea, as well as the landmasses that make up southern Europe, the Middle East, and North Africa. The shoving together (compression) created many of the Mediterranean mountain ranges, including the Apennines, the Dinaro-Hellenic chain of the Balkan States, the Alps, and the Pyrenees. The sections being pulled apart (extension) have become ocean basins, including the Tyrrhenian Sea, part of the Aegean Sea, and the large Liguro-Provençal Basin between Sardinia and Spain.

The most visible product of all this pushing and pulling of the Earth's crust—the Mediterranean Sea—became the core of the Roman Empire. In addition to being an obvious source of food, the Mediterranean was (and continues to be) the main thoroughfare for merchant and military ships. It provided the access needed to develop an extensive empire and the protection needed to defend that empire from other powers.

When we examine stones used by the Romans for both art and building materials, we can see both the geographic extent and the geologic scope of the Empire. Marble came from Italy, the Balkan Peninsula, Turkey, Egypt, France, Asia Minor, and Spain; granite, gabbro, serpentine, and porphyry were brought back from Egypt; and alabaster was returned from Egypt, Tunisia, and Algeria. Roman artwork used precious metals and gemstones from all corners of the Empire.

This preview of Rome's geologic and geographic setting, taken in seven steps, also is a preview of this book, revealing the constant interaction between geologic setting, humans, urban development, and people's response to the natural phenomena that have affected the city.

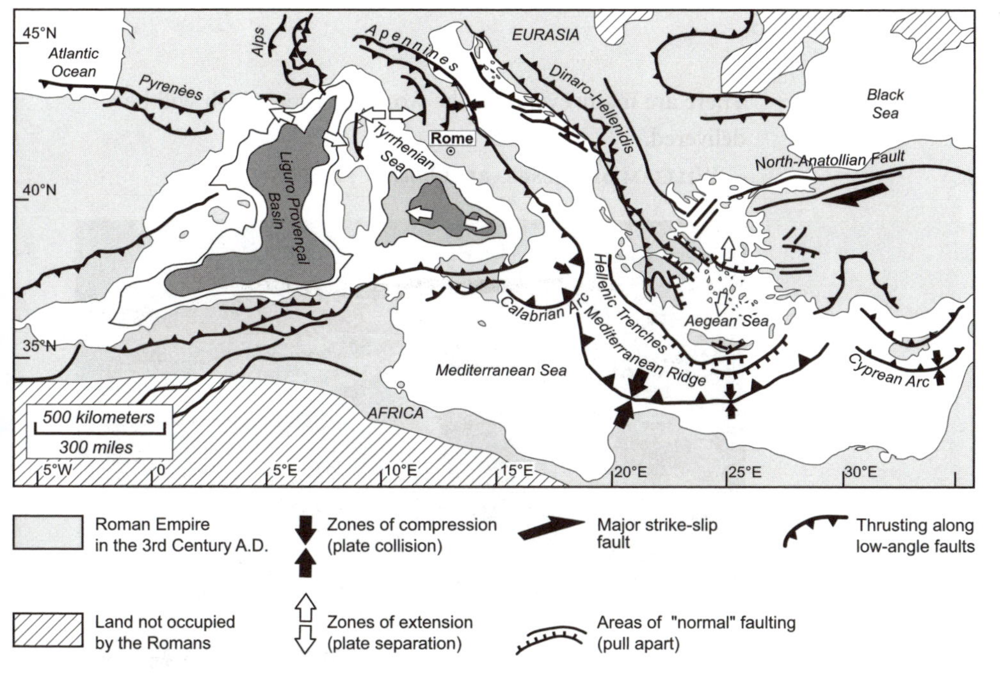

In the last window, we can see the extent of the later Roman Empire (at the time of Diocletian in the 3rd century A.D.) on a simplified geologic map that depicts the Mediterranean Sea and the landmasses around its margins. Our window is a bit distorted here, with the 3,600-kilometer measurement in an east-west direction to encompass the Mediterranean. Complex plate tectonic processes formed the seas that Rome found so crucial for military and merchant transportation. This geologic setting offered up a rich variety of metals, gemstones, and building stone used by the Romans, as well as a variety of soils required for growing wheat, grapes, and olives for oil.

The area's complex geologic pattern was shaped by movements of the African and Eurasian crustal plates. Compressive and extensional forces in this zone are overwhelmingly complicated, even to geologists. Compression (shoving) has produced the thrust faults (marked with "teeth" and dark arrows), and extension (pulling apart) has opened up basins in the sea and on land, such as the deep Liguro-Provençal basin, located between the Spanish coast and Sardinia. Volcanic activity occurs throughout and includes the many Italian volcanoes on Sicily, in the Tyrrhenian Sea, and along the western coast of the Italian Peninsula. (From a diagram by Claudio Faccena, University of Roma Tre)

TIMELINES

There are many events in the womb of time which will be delivered.
—William Shakespeare, *Othello*

Archeological excavations in the Roman fora unravel the last 3,000 years of Rome's history one layer at a time, as shown in this photograph of a dig in the Forum. It is slow, meticulous work that requires patience, experience, and a good eye. To understand the geologic history of Rome, we must go backward 300 million years, as summarized in the timelines presented here.

The links between the history of Rome and its geologic foundation are strong, yet they are rarely considered, in part because of philosophical differences between historians and geologists. Both professions deal with history, but in very different ways. Historians focus on the interactions between people, primarily within a culture but also between cultures; their approach is usually linear: events are located on timelines that span years, decades, and centuries; and their work is based chiefly on the written word, but occasionally on oral traditions as well. Geologists are concerned with interactions between natural elements and, increasingly, between nature and humans. They must rely on the mute witness of rock and soil. Finally, given the vastness of geologic time, geologists tend to think in "powers of ten." We begin this book by examining the Trevi Fountain and then rise above it in steps of ten, initially looking down on a piazza about 30 meters on each side and finally viewing a vast region 3,000 kilometers on a side. In the timelines presented here, we begin our journey 300 million years ago and move toward the present in similar steps—but in time rather than distance. The geologic processes that have shaped the Rome and surrounding region we see today are part of a continuum, but to evaluate them in a linear fashion would require more time and paper than we could envision for any book!

Italy in general and the Apennines in particular have some of the most complex geology in the world. The effort to understand how the Italian Peninsula and the Apennine chain were formed has required detailed fieldwork by hundreds of geologists over the last hundred years. Scientific concepts and technologies have changed, but the rocks are the same, as are the locations in which they were described by pioneering geologists—so even the most rapid of plate tectonic movements haven't changed the shape that much. In the late 1970s and early 1980s, the Italian government began the Geodynamics Project to integrate all that was known about the geology of Italy to better understand the nation's natural hazards and geologic resources. To assemble the geologic map of Italy in five years required the efforts of nearly 100 geologists (not counting students), coordinated by Maurizio Parotto of the University of Rome. Geologists often disagree, usually in a friendly, competitive way, about the structure, rock types, and ages of units; thus, compiling a single geologic map of Italy also required considerable strength and diplomacy. Field, laboratory, and numerical studies of geologic problems of the Apennines continue to constantly refine and update what is known, especially in the light of new ideas concerning the dynamic Earth.

CHAPTER ONE

300 Million to 20 Million Years before the Present

Most of the rocks in the Apennines in central Italy are varieties of limestone (calcium carbonate) or dolomite (calcium magnesian carbonate), with some sandstones and claystones. Nearly all these rocks were deposited as sediments in shallow to intermediate-depth seas that covered the region from about 300 million to 10 million years ago (timelines 1 and 2). Until about 15 million years ago, they were part of a broad seaway (the Tethys Sea), which spanned much of what is now Europe and connected the Indian and Atlantic oceans. These ocean basins were not just static areas where carbonate mud accumulated for hundreds of millions of years; they were dynamic, shifting at rates of a few centimeters (about an inch) per year as the Earth's crust moved. To understand these marine environments of the past, we need to pull apart all the crumpled and stacked layers of rock—at least on paper. Building on their lifetimes of work in the Apennines, Antonio Praturlon and Maurizio Parotto of the University of Roma Tre have done just that and have given us glimpses of the early seas, islands, and basins that occupied the region.

A number of tectonic, volcanic, and sedimentary events were occurring at the same time. For example, from about 220 to about 70 million years ago, there was a large oceanic carbonate platform across what now is Lazio and Umbria. (To envision this environment, think of the present-day Bahamas and the shallow seas that surround them, all of which are actually situated on a platform that rises above the floor of the Atlantic Ocean.) The ancient carbonate platform of Lazio and Umbria rose above a sea floor that was in motion. The water depth and depositional environment changed with time: about 220 to 190 million years ago, the ocean floor began to break up; some blocks rose above sea level, and others were submerged along down-dropped sections. The resulting environments then included areas from above sea level and from adjacent oceanic basins that had slowly filled with limy muds and sands, as well as some reefs that may have been present in the shallower seas. Small ocean basins inset into the platform were also filled with lime mud and sands that eventually became laminated limestones. These geologic conditions existed until about 19 million years

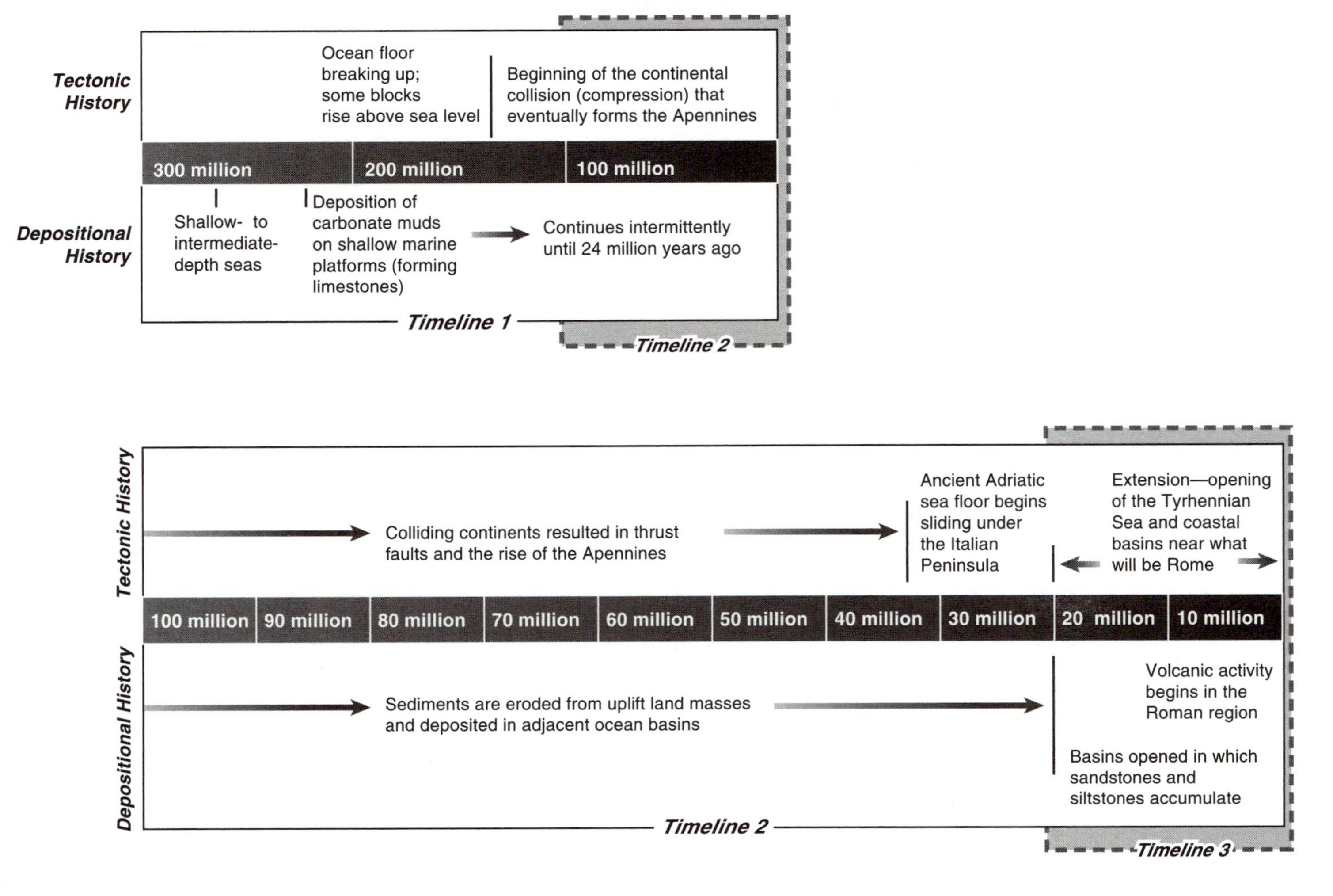
Tectonic History
Ocean floor breaking up; some blocks rise above sea level
Beginning of the continental collision (compression) that eventually forms the Apennines
300 million
200 million
100 million
Depositional History
Shallow- to intermediate-depth seas
Deposition of carbonate muds on shallow marine platforms (forming limestones)
Continues intermittently until 24 million years ago
Timeline 1
Timeline 2
Tectonic History
Colliding continents resulted in thrust faults and the rise of the Apennines
Ancient Adriatic sea floor begins sliding under the Italian Peninsula
Extension—opening of the Tyrhennian Sea and coastal basins near what will be Rome
100 million
90 million
80 million
70 million
60 million
50 million
40 million
30 million
20 million
10 million
Depositional History
Sediments are eroded from uplift land masses and deposited in adjacent ocean basins
Volcanic activity begins in the Roman region
Basins opened in which sandstones and siltstones accumulate
Timeline 2
Timeline 3

ago. The sediments, eventually turned into marls (a carbonate-rich claystone) and mudstones, are now exposed in the Apennines of Umbria and Marche to the north and Sicily to the south. Evidence that could reveal the extent of the ocean basins adjacent to the Lazio-Umbria carbonate platform has been more or less obliterated by later continental collisions.

About 130 to 33 million years ago, plate collisions compressed marine sedimentary rocks along a north-south line (see timeline 2). Instead of bending like a rug, the layers that made up the upper part of the Earth's crust, including the sedimentary rocks, broke along myriad north-south-trending, low-angle faults that slid one over the other along very low, westward slopes. Imagine a sheet of thick cardboard lying on a table. You cut completely through it along a number of parallel lines, which we'll use as an analog to faults. Now, while you hold the right side motionless, you push the left side toward the right, which compresses the cut sections and stacks then up so they slant downward toward your left hand. This stack of cardboard sheets is the equivalent of the early Apennine mountain chain. As you can see if you visit the high Apennines around Gran Sasso, northeast of Rome, the reality is much more complex, but this simulation gives you a general idea of what has happened over the last hundred million years.

20 Million Years Ago to the Present

In contrast to the compression that for hundreds of millions of years had stacked up sedimentary rocks to form mountains, the Earth's crust in this part of the world now began to pull apart and stretch (timelines 3–7). About 25 to 20 million years ago, the thinned crust below what is now the Adriatic Sea began sliding under the Italian Peninsula to form what is called a *subduction zone.* At the same time, the upper portions of the western Mediterranean, the Ligurian Sea, began to open as the crustal blocks we now know as Sardinia and Corsica drifted to the east with a counterclockwise motion. Collapse of the crumpled wedge that was the Alps pushed deeply buried rocks to the surface (one of the best examples of these rocks is the Carrara marble so frequently used in the great sculptures of Italy). Stretching and thinning of the

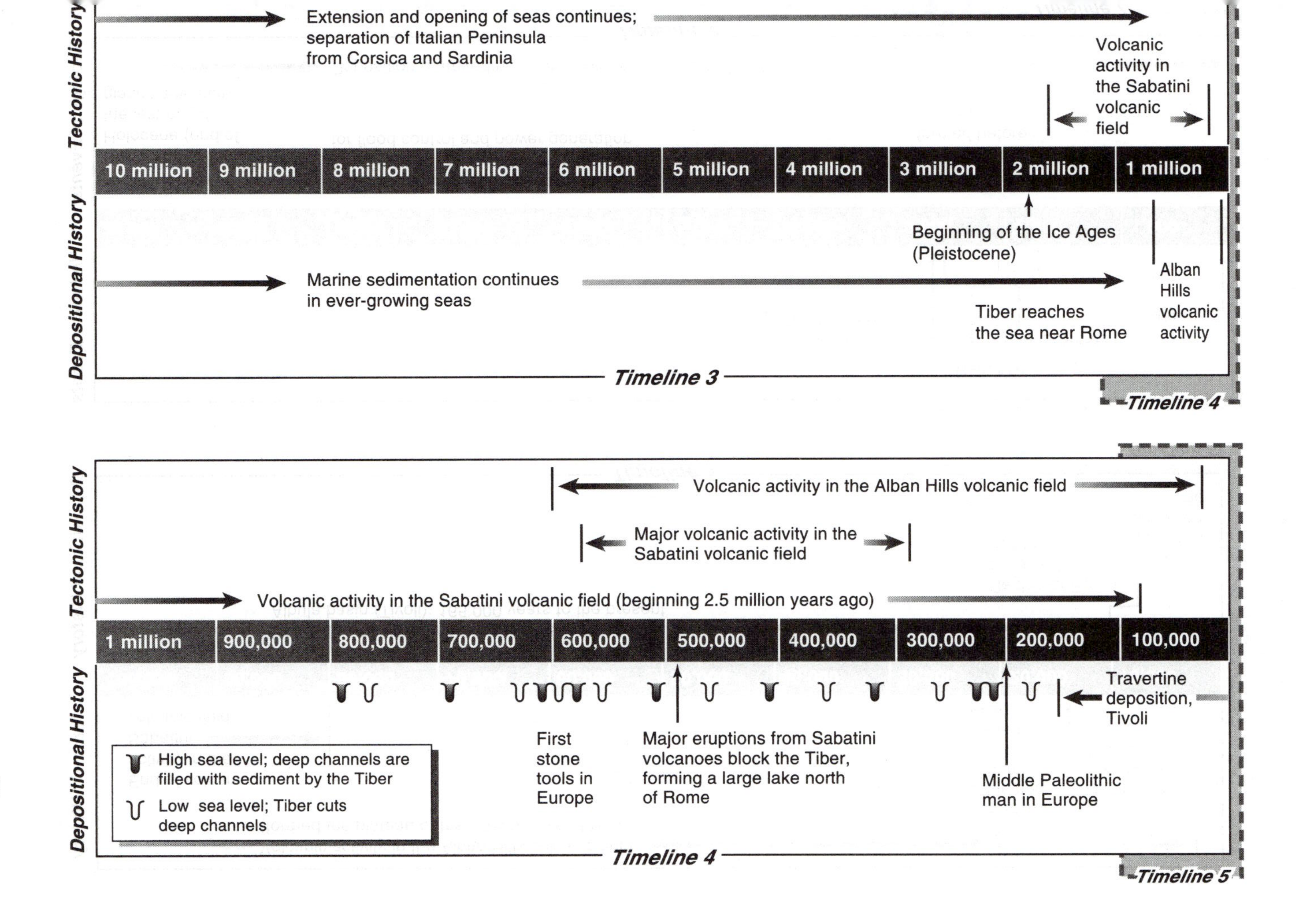
Tectonic History
Extension and opening of seas continues; separation of Italian Peninsula from Corsica and Sardinia
Volcanic activity in the Sabatini volcanic field
10 million
9 million
8 million
7 million
6 million
5 million
4 million
3 million
2 million
1 million
Depositional History
Beginning of the Ice Ages (Pleistocene)
Marine sedimentation continues in ever-growing seas
Tiber reaches the sea near Rome
Alban Hills volcanic activity
Timeline 3
Timeline 4
Tectonic History
Volcanic activity in the Alban Hills volcanic field
Major volcanic activity in the Sabatini volcanic field
Volcanic activity in the Sabatini volcanic field (beginning 2.5 million years ago)
1 million
900,000
800,000
700,000
600,000
500,000
400,000
300,000
200,000
100,000
Depositional History
Travertine deposition, Tivoli
First stone tools in Europe
Major eruptions from Sabatini volcanoes block the Tiber, forming a large lake north of Rome
Middle Paleolithic man in Europe
High sea level; deep channels are filled with sediment by the Tiber
Low sea level; Tiber cuts deep channels
Timeline 4
Timeline 5

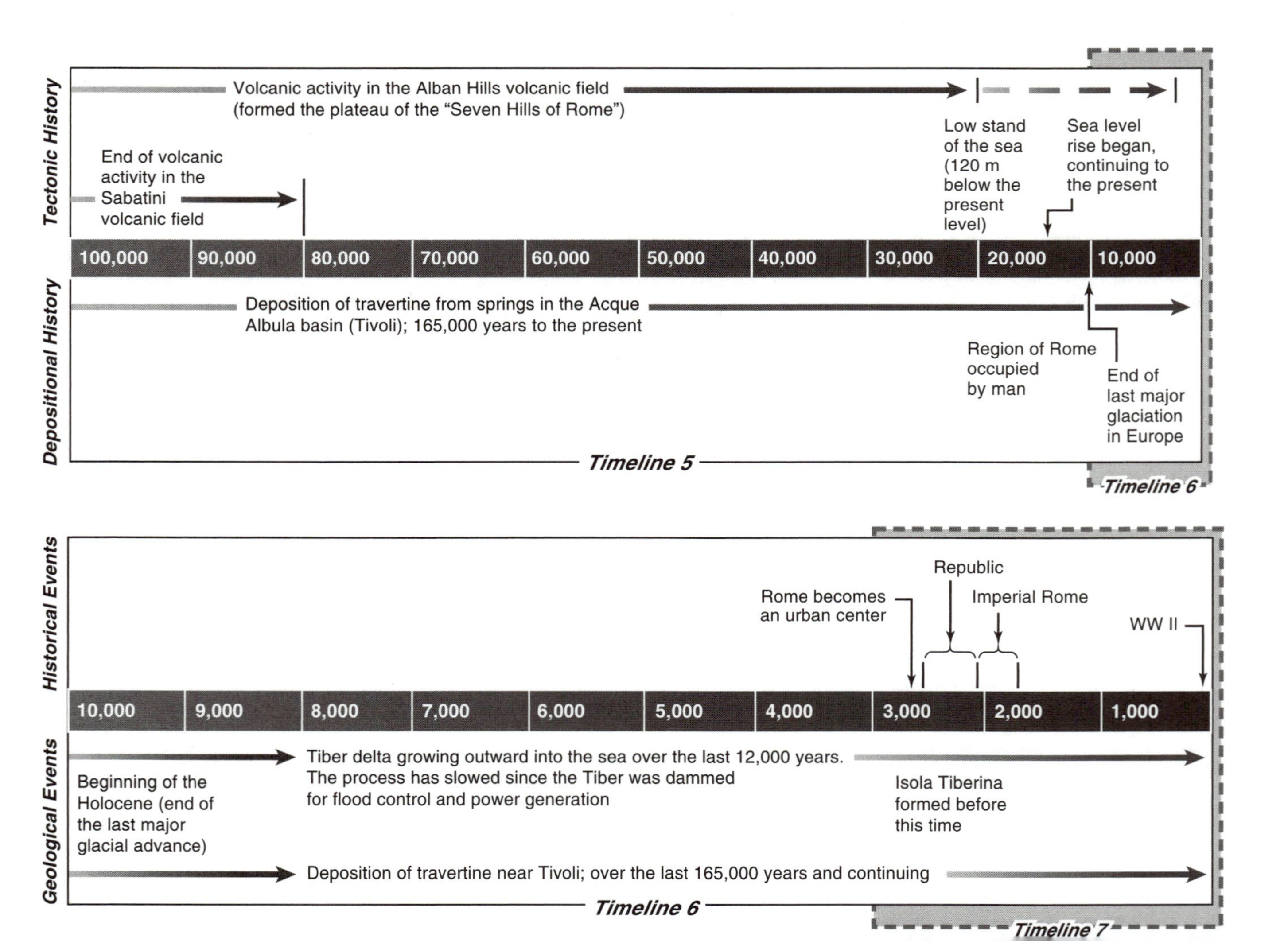
Tectonic History
Volcanic activity in the Alban Hills volcanic field
(formed the plateau of the "Seven Hills of Rome")
End of volcanic activity in the Sabatini volcanic field
Low stand of the sea (120 m below the present level)
Sea level rise began, continuing to the present
100,000
90,000
80,000
70,000
60,000
50,000
40,000
30,000
20,000
10,000
Depositional History
Deposition of travertine from springs in the Acque Albula basin (Tivoli); 165,000 years to the present
Region of Rome occupied by man
End of last major glaciation in Europe
Timeline 5
Timeline 6
Historical Events
Rome becomes an urban center
Republic
Imperial Rome
WW II
10,000
9,000
8,000
7,000
6,000
5,000
4,000
3,000
2,000
1,000
Geological Events
Tiber delta growing outward into the sea over the last 12,000 years.
The process has slowed since the Tiber was dammed
for flood control and power generation
Beginning of the Holocene (end of the last major glacial advance)
Isola Tiberina formed before this time
Deposition of travertine near Tivoli; over the last 165,000 years and continuing
Timeline 6
Timeline 7

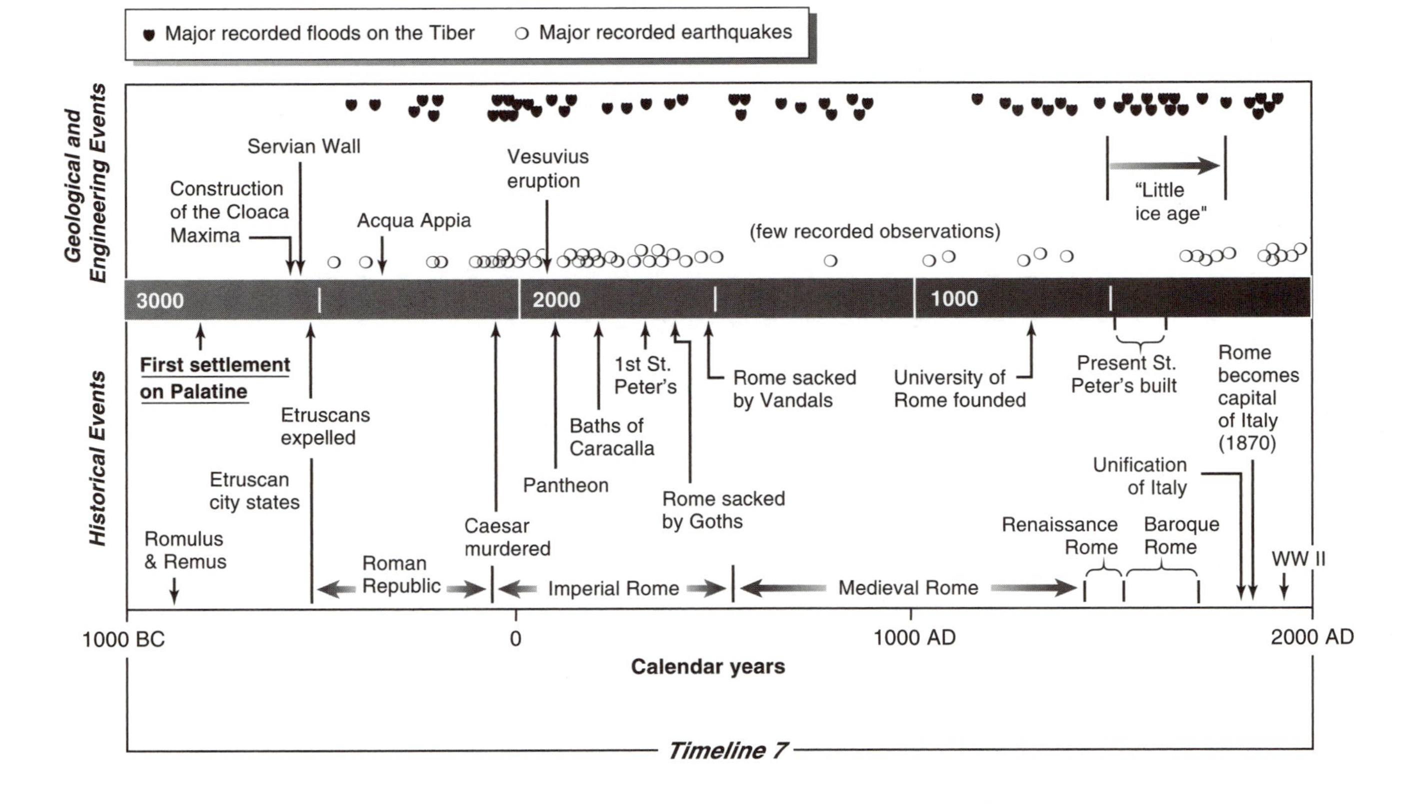
Major recorded floods on the Tiber
Major recorded earthquakes
Geological and Engineering Events
Historical Events
Servian Wall
Construction of the Cloaca Maxima
Acqua Appia
Vesuvius eruption
(few recorded observations)
"Little ice age"
3000
2000
1000
First settlement on Palatine
Etruscans expelled
Etruscan city states
Romulus & Remus
Roman Republic
Caesar murdered
Pantheon
Baths of Caracalla
1st St. Peter's
Rome sacked by Goths
Rome sacked by Vandals
Imperial Rome
Medieval Rome
University of Rome founded
Present St. Peter's built
Renaissance Rome
Baroque Rome
Unification of Italy
Rome becomes capital of Italy (1870)
WW II
1000 BC
0
1000 AD
2000 AD
Calendar years

Timeline 7

crust also produced a setting for the beginnings of volcanism, an activity that continued intermittently from about 600,000 to 3,500 years ago in volcanic fields that flank Rome to the north and south.

The pulling apart of the Tyrrhenian coastal plain over the last 20 million years had formed elongate basins up to 10 kilometers (6 miles) wide and 2,500 meters (8,200 feet) deep, bounded by "normal" faults (in which one side moves down relative to the other) that are more or less parallel to the coast. The subsiding basins were rapidly filled with debris washed down from the Apennines or from the volcanoes that were erupting during the same period. Because they were close to the coast, sediments were deposited both above and below sea level, thus getting a sequence of interbedded marine and "continental" sedimentary rocks in the area of Rome. Regardless of the type of motion (pushing or pulling), the structural trends of features such as the Apennine mountain chain or coastal basins have stayed more or less the same over the last several hundred million years.

The depth of the Tiber River valley has changed a great deal over the last 13,000 years. The river quickly eroded its valley and tributaries to develop equilibrium relative to the sea, which was once nearly 70 meters (230 feet) below its present level. As sea level rose, the Tiber stopped eroding and began depositing sediment, which is now quite thick below the Tiber as it passes through Rome.

Each of the major events that occurred during this timeline of 300 million years took tens of millions to millions of years—with two exceptions: the arrival of humans 600,000 years ago (just 0.2 percent of the timescale) and the founding of Rome 2,800 years ago (a mere 0.0006 percent of the timescale). Despite the ways we have changed the face of the planet, *Homo sapiens* hasn't been around for long.

CHAPTER 2

Center of the Western World

THE CAPITOLINE (CAMPIDOGLIO) HILL

THE CAPITOLINE HILL is one of the most-photographed hills in the world, although most camera-bearing tourists don't realize the significance of this promontory behind the Roman Forum. Many of the large brown blocks of tuff (consolidated volcanic ash) used to construct the Forum were excavated from the flanks of this hill, but the summit itself was and remains one of importance. This small plateau (about 0.1 square kilometers or 24 acres) was the center of power and religion for what was Western civilization 2,000 years ago. Among the seven hills of Rome, the Capitoline is one of three nearest the Tiber (Capitoline, Aventine, and Palatine) that were important for both defense and access to the river so crucial to Roman commerce.

There is much to see on these 24 acres. Go south from the Piazza Venezia along the Via del Teatro di Marcello and look up through tree-lined slopes to the relatively modern palazzos (now mostly museums) on the summit, where decisions were once made that affected the entire Western world. There are many optional routes here.

If you choose to go northwest, you can approach the plateau topped with the Piazza del Campidoglio via the ramp designed by Michelangelo (the Cordonata) in 1559–66 (go ahead—feel important!).

If, instead, you continue along the Via del Teatro di Marcello, around the hill at the Via Jugaro to the Via di Monte Caprino and up the steps toward the Via del Tempio di Giove, you will be circumnavigating the Tarpeian Rock, a cliff along the southern edge of the plateau that was used as a platform for throwing traitors to their death.

Today the plateau summit includes the Piazza del Campidoglio, designed by Michelangelo, and buildings on three sides: the Palazzo Senatorio (Rome's Town Hall), the Palazzo dei Conservatori, and the Palazzo Nuovo (the last two are now museums worth visiting).

On the northern slope of the Capitoline, above the busy Piazza Venezia, is the Vittorio Emanuele Monument, an easily identified limestone landmark that was built between 1885 and 1911 to honor the first king of a unified Italy.

The Capitoline Hill is one of the smallest of Rome's seven hills and consists of three geologic units produced by activity over the last 600,000 years: sands and lake deposits from the floodplain of the ancient Tiber, interbeds of travertine left by mineral springs, and pyroclastic flow deposits (tuffs) erupted from volcanoes in the Alban Hills as well as a carapace of river and lake sediments. As is the case with the other six hills, this promontory is held up mainly by the tuffs left by eruptions in the Alban Hills volcanic field, southeast of Rome.

You may ask, "Volcanoes in Rome?"

Volcanoes—In Rome?

The "grand tour" of Europe during the 18th and 19th centuries invariably included a visit to Vesuvius. Great artists and writers, including Joseph Wright of Derby, Johann Wolfgang von Goethe, and Mark Twain, were enthralled by then-active Vesuvius's eruptions and quiet contemplation of the historic last days of Pompeii. The volcanoes of the Phlegrean Fields, Etna, and Stromboli—all farther south—were often cited by both pagan and Christian theologians as possible entrances to the underworld. In fact, the science of volcanology began in Italy with observations of Vesuvius and Mount Etna.

Today's tourists commonly include visits to at least one of Italy's well-known volcanoes on their itineraries . . . but volcanoes near Rome *itself*? Most tourists and even many Romans are unaware that the city is flanked by two very large volcanic fields: the Alban Hills to the southeast and the Sabatini volcanic field to the northwest. The seven hills of Rome owe their very existence to the volcanoes of the Alban Hills. The threat of volcanic eruptions is not imminent, but these volcanic fields are "geologically young" (less than a million years old).

Looking southeast from the Capitoline summit in the park near the Temple of Jupiter, you can see the volcanic Alban Hills (Colli Albani), site of the "Castelli Romani" and the pope's summer palace at Castel

This digital elevation map provides an excellent view of the basic structures and the relationships between the Latium volcanoes (the Alban Hills and the Sabatini volcanic fields), the Apennines, and the Tiber River delta.

Gandolfo. The massif visible from the Capitoline is an accumulation of overlapping deposits from large and small eruptions from more than seventy-two craters that were active intermittently over the last 600,000 years. The tuffs and lava flows of the Alban Hills cover an area of about 1,600 square kilometers (633 square miles); on a clear day, you can see their inverted shieldlike outline from the Capitoline or the Janiculum Hill or by looking southeast along the Circus Maximus.

A Brief History of the Alban Hills

Approximately 50 kilometers (31 miles) in diameter, the Alban Hills span the coastal plain between the Apennines and the sea. The hills rise to an elevation of nearly 1,000 meters (3,300 feet) above sea level, where

The Alban Hills volcanic field, as seen from Monteverde above the Trastevere neighborhood in eastern Rome. The summit of the Alban Hills consists of a wide collapse crater (caldera) and irregular high ground, reflecting smaller volcanoes erupted after the caldera collapse (see the previous figure for a map view).

the broad summit is dominated by a caldera (collapse crater) that is now mostly buried by material from younger volcanoes. The crater lakes of Albano and Nemi (Lacus Albanus and Lacus Nemorensis) provide the most obvious evidence of past volcanic eruptions and can be visited during a day trip from Rome.

In ancient times, the Alban Hills were more heavily vegetated and exhibited a greater diversity of plants than we see today. The forested slopes were once covered with oak, hazel, and maple trees, but only rare patches of the mixed forest now remain in the Chigi and Arriccia parks. The chestnut tree (a symbol of the Campagna Romana) has replaced most of the mixed forest, having been introduced to the area in the 17th century.

Much of the volcanic field is now covered with small towns, villas, monasteries, and vineyards. Archeological evidence from the Alban

Lago di Albano (Lake Albano) is one of the youngest volcanoes on the slopes of the Alban Hills volcanic field. The papal summer palace is located on the rim of this deep crater, which was formed during multiple eruptions of overlapping volcanoes. The youngest (Late Bronze Age) volcanic deposits were erupted from Lake Albano and reach the edge of modern Rome.

Hills, particularly around the edges of the Nemi and Albano lakes, indicates that humans have occupied the area since the Bronze Age. Such early sites, including some of the many scattered villages across the Campagna Romana, provided a framework for the development of Roman civilization. Historic and prehistoric sites within the Alban Hills are testimony to the presence of countless generations who were attracted to this gentle volcanic terrain, with its excellent water sources and fertile soils.

The plateau that slopes from the summit of the Alban Hills into Rome allowed the city to gradually expand upward and outward into the countryside to the southeast. There is historical evidence that the ancient Latin settlement of Alba Longa, along the margin of Lake Albano, has always had strong connections with Rome. In contrast, Rome was slow to develop to the north and west because of the natural barriers of the Tiber and the steep slopes of Monte Mario, which restricted communication and transport toward Lake Bracciano.

Can we forecast an eruption in the Alban Hills? Probably not. The Alban Hills' volcanic activity peaked during three phases, between 600,000 and 300,000, between 300,000 and 200,000, and again between 200,000 and 20,000 years ago. To most people, this seems long ago, but to geologists it was "just yesterday." Enough time passed between the periods of volcanic activity for six soils to have been formed on ancient slopes before being buried by later eruptions; these ancient soils are now visible as "marker beds" that separate the volcanic ash deposits and lavas of each eruption phase. One of this book's authors (De Rita) has observed that the periods of volcanic activity occurred during times of lower sea level (during the ice ages) and that volcanic activity may have been stimulated by lowered pressure in magma (molten rock) chambers underlying the volcanoes. Volumes of erupted ash and lava decreased steadily with time, from 283 cubic kilometers (68 cubic miles) during the first eruption phase to 6 cubic kilometers (1.4 cubic miles) during the second, and to 1 cubic kilometer (0.25 cubic mile) during the third. This may indicate waning activity, but we don't really know if these volcanoes are extinct. To determine the risk related to volcanic eruptions is difficult, and the significance of these numbers certainly has not decreased real estate values!

Deposits from the Alban Hills' earliest eruptions are now buried under younger volcanoes, so we don't have a complete record of the earliest volcanic activity. As far as we know, 98 percent of the tuffs and lavas that exist today in the Alban Hills were erupted between 600,000 and 300,000 years ago. Most of the volcanic field's deposits were produced by pyroclastic flows (mixtures of hot volcanic ash and gases) that moved in all directions for tens of kilometers and into what is now Rome. The largest eruptions—nearly all of which occurred when rising magma mixed with groundwater in deep aquifers (below the coastal plain)—produced pyroclastic flows that rushed 30 to 50 kilometers (19 to 31 miles) from their craters at hurricane velocities, covering collectively about 1,000 square kilometers (386 square miles). Consolidated volcanic ash deposits left by the pyroclastic flows progressively modified the drainage network that flowed into the ancient Tiber by clogging valleys around the volcanoes. The pyroclastic flows that reached what is now Rome buried the ancient Tiber River, forcing it to the west. After these catastrophic eruptions, a 10- to 12-kilometer-diameter caldera dominated the summit of the volcanic field. The Tuscolano-Artemisio caldera may have gradually subsided in a piecemeal manner before being filled with ash and lava. Lava flows that erupted during this period are 30 meters (98 feet) thick at the caldera's rim.

Eruptions at vents within the summit caldera continued intermittently over the period between 300,000 and 200,000 years ago. The products of these eruptions include the Faete cone and, more notably, the Capo di Bove lava, which flowed down a northwest-trending valley toward the ancient Tiber, creating a sloping plateau later followed by the ancient Appian Way. Much of the stone used to surface the durable highways near Rome came from this lava flow.

The most recent crater-forming eruptions of the Alban Hills occurred between 200,000 and 20,000 years ago at the craters of Pantano Secco, Prata Porci, Valle Marciana, Castiglione, Nemi, Arricia, Giuturna, and Albano. The largest of these, Lake Albano, consists of five overlapping craters and is flooded. The 3.5-by-2.4-kilometer (2.2-by-1.5-mile) crater is 442 meters (1,450 feet) deep, with a water depth of 175 meters (574 feet).

Geologic and archeological studies in 2001 and 2002 have changed our views of the volcanic risks posed by the Alban Hills. Civil works excavations on the Ciampino plain (immediately southeast of Rome, just outside the ring road) revealed two previously unknown explosive volcanic deposits and several volcanic mudflow deposits. Eruption activity from the Lake Albano crater is much younger than was previously believed, extending well into the Holocene (the last 10,000 years). Catastrophic overflows from the lowest rim of the lake left significant mudflows—as recently as the 4th century B.C.—that formed the smooth surface of the Ciampino plain. The overflows may have been triggered by sudden injections in the lake bottom of carbon dioxide—rich fluids that are present underneath the volcano even today. High carbon dioxide releases continue to be a hazard in the Alban Hills, especially in residential areas and vineyards near Ciampino Airport and Bagni di Tivoli.

Strong to moderate discharges of both carbon dioxide and hydrogen sulfide, commonly associated with volcanic fields, increase when there are earthquakes; this phenomenon has been noted here at least fifteen times during the last 200 years. It's possible that the greatest threat posed by the Alban Hills at this time is the emission of carbon dioxide, which, because it is heavier than air, can pool in depressions such as craters and basements. When there is no wind, the gas can be a severe hazard, suffocating any form of animal life unlucky enough to enter one of these depressions.

Lake Albano's water level can rise when carbon dioxide accumulates in the bottom of the lake, but the depth can also vary as a result of rainfall or fluctuation in withdrawal by water utilities. Early Roman engineers kept the water level constant at 293 meters above sea level by means of an underground drain that may have been excavated as early as the 4th century B.C. For unexplained reasons, the lake level has been dropping recently, exposing old beaches and Roman ruins.

Lake Nemi also has a drain that dates to the 4th century B.C. The lake was drained entirely between 1927 and 1932 to salvage two Roman ships that were built for Caligula and used for holiday celebrations and religious ceremonies; the ships were destroyed by a fire in 1944. Unlike Lake Albano, Lake Nemi is now heavily polluted.

The best vineyards in the Alban Hills are located in the broad floors of the craters of Valle Marciana, Pantano Secco, and Prata Porci in the western part of the volcanic field. Here the local grapes (varieties of Malvasia), grown on American rootstock, produce a dry white wine. The craters are excellent sites because the deposits are very permeable and the soils are rich in potassium. Vineyards located on the tuff plateaus or in crater bottoms allow easy harvesting, but in the past the vineyards were also located on crater sides, where they were harvested by hand. The best exposures face south, southwest, and northwest. The combination of volcanic soils and microclimate produce excellent straw-yellow, dry wines; these "golden wines" from the Alban Hills have been popular since before Imperial Roman times to the present.

Is Rome Threatened by Volcanic Eruptions?

As you stand at your viewpoint on the Capitoline Hill, do you need to worry about being wiped out by an eruption? We don't know. Nothing catastrophic has happened here for thousands of years, but the processes that shape our planet work on a timescale that is very different from that of human history. The famous eruption of Vesuvius in A.D. 79 followed a long period of inactivity—long enough that neither the Greeks nor the Romans knew that Vesuvius was a volcano. There have been recent swarms of small earthquakes below the Alban Hills, with focus depths of 4 to 6 kilometers (2.5 to 3.7 miles), which have been interpreted as energy releases from a cooling magma chamber (note an emphasis on "interpreted"). One of these earthquake swarms occurred from 1989 to 1990, with up to forty small earthquakes each day. As you read earlier in this chapter, there is also a possible risk from massive eruptions of carbon dioxide from Lake Albano.

Ancient Romans were very concerned about portents, and extreme attention was paid to any unusual natural event. Livy, Pliny the Elder, Orosius, and Julius Obsequens described interesting groups of portents, including "rains of stones together with babies," which, considering the duration of the event and the atmospheric and acoustic phenomena, were interpreted as volcanic. There may be a link between "rains of stones" and the locations of recent gas emissions, especially an explosive release of carbon dioxide. Another interpretation of such reported events is that they could have been gas emissions that accompanied the collapse of a sinkhole. A similar historic event occurred in the Sabatini volcanic field near Monte Soratte, where gas was released and blew rocks out of a hole in the small Lago Puzzo crater.

Some events described by ancient Roman writers could have been phreatic eruptions (steam blasts that take place when pressure within a geothermal system exceeds that of the overlying rock). Phreatic eruptions occur in many parts of the world in association with volcanic eruptions, sometimes far from any volcanoes with recent activity, and even when drillers lose control of a steam well, causing a "blowout." Phreatic eruptions occurred in the Roccamonfina area about 290 B.C., and there is evidence of historic phreatic blasts throughout the Phlegrean Fields near Naples.

Both Julius Obsequens and Pliny the Elder reported that the Alban Hills burned at night in 129 B.C. However, there is no verification of any volcanic activity in the Alban Hills at that time—this event was more likely a forest fire or brushfire.

• • •

You can descend from the Capitoline Hill from the park near the Temple of Jupiter or from the Piazza del Campidoglio by way of the Via del Tarpeo to the Via della Consolazione. Below the eastern edge, along the Via della Consolazione, you actually see outcrops of the tuffs that make up this famous hill—a sight guaranteed to make a geologist's pulse race in this city that is mostly buried by man-made debris and buildings! When you finish your visit to the Forum, climb to the verdant plateau of the Palatine Hill, which has been occupied since the Iron Age and by tradition is where Romulus founded the City of Rome in 753 B.C.

CHAPTER **3**

Palaces and Gardens

THE PALATINE (PALATINO) HILL

. . . but were I Brutus,
And Brutus Antony, there were an Antony
Would ruffle up your spirits and put a tongue
In every wound of Caesar that should move
The stones of Rome to rise and mutiny.
—William Shakespeare, *Julius Caesar*

The Palatine Hill is evident from all sides, its prominent, tablelike form covered with ruins and trees. One of Rome's top attractions, the Palatine is believed to be the first of Rome's seven hills to be inhabited—and perhaps the original nucleus from which the great city evolved. It was strategically located close to the Tiber, yet high enough for defense and a good breeze on hot summer days. To see the overall form of this small plateau, start at the northwest end of the Circus Maximus and walk southeast. This allows a view of the end of this rectangular plateau and its ruins. When you reach the end of the circus, turn left up the Via di San Gregorio and walk along the eastern margin of the hill toward the Colosseum. This will take you to the entrance to the Roman Forum and your access to the Palatine Hill. One ticket pays for your entry to both sites. Follow signs to various access points (the steps at Atrium Vestae, Santa Maria Antiqua, or the Clivus Palatinus road).

The Palatine is composed of rocks very similar to the sequence that underlies the Capitoline: a series of pyroclastic flow deposits (tuffs) from the Alban Hills that overlie river and marsh deposits. Overlying the tuffs is a veneer of sediment from younger rivers and adjacent marshes. The more or less tabular stack of deposits was later incised by streams flowing from the plateau into the Tiber, leaving the present tuff plateau as an erosional remnant.

The Palatine is a quiet place, with pleasant breezes and surrounded by cliffs or steep slopes. You can imagine why this 44-acre plateau was a desirable location, perhaps from the earliest settlements in what is now Rome into the 16th century, when it was home of the Farnese Gardens built by Cardinal Alessandro Farnese. Although most of the later Imperial palaces of Augustus, Tiberius, Caligula, and Domitian are constructed of brick, the earliest stone buildings, including the Temple of Cybele, were constructed of tuff from the plateau itself.

What Are Tuffs?

The term *tuff* is derived from *tufo*, a word originally used by Italian quarrymen for a rock that can be cut with a knife. *Tuff* is used by geologists to designate consolidated deposits of ash, pumice, and rock fragments left behind by explosive volcanic eruptions, similar to the material underling much of the Capitoline and Palatine hills. A variety of tuffs have been used throughout the history of Rome. In guidebooks and art books describing the architecture and archeology of Rome, you'll see a bewildering number of names applied to these rocks—a nomenclature that has evolved over the past 2,000 years. The most confusing of these terms is *tufa*, sometimes used in place of *tuff*. Within the world of geologists, these terms refer to two very different rocks. The first, *tufa*, is actually a carbonate spring deposit, whereas *tuff* refers to a consolidated volcanic ash deposit. Even more confusingly, *tuff* in Italian is *tufo*. So, to test your understanding, which are the correct words for the consolidated volcanic ash deposits left by explosive volcanic eruptions—*tuff*, *tufa*, or *tufo*?

The ideal tuffs for quarrying are those from volcanoes that quickly erupted large volumes of volcanic ash particles and gas; the ash and gas sped across the countryside as density currents (pyroclastic flows), burying the land with thick ash deposits within minutes or hours. The consolidated deposits left by pyroclastic flows are called *ignimbrites*. The large volcanoes located north and south of Rome are somewhat unique because of the type of explosive eruptions that shaped their craters and produced the deposits on their flanks. The usual energy of explosive volcanic fragmentation was augmented when rising magmas encountered groundwater in limestone aquifers that underlie the coastal plain near Rome (these limestones correlate with those exposed in the Apennines). Such an eruption phenomenon, referred to as *hydrovolcanic*, occurs when very hot (magma) and cold (water) fluids mix. The resulting explosions are similar to the phenomenon so feared within foundries, where even a small amount of water in the bottom of a mold can lead to catastrophic explosions as the mold is filled with molten metal. Pyroclastic flow deposits left by hydrovolcanic eruptions consist of extremely fine-grained, glassy ash particles that are chemically unstable when deposited and are subsequently altered by water and steam. The minerals formed during the glass-water interaction bind the ash particles, thus creating the durable stone that was—and still is—quarried in and around Rome.

Another, very dense, type of tuff consists of ash and pumice particles that are actually welded by the residual heat of the deposit within weeks or months after deposition—this rock is called a *welded tuff*. In these tuffs, *fiamme* (black, flame-shaped pumice fragments flattened and welded by the weight of the overlying hot deposit) are disseminated through the rock. Welded tuffs are not common around *either* of the Roman volcanic fields, but can be found farther north, near Cimino and Viterbo, where they are quarried for use throughout central Italy as architectural trim and sculpture.

Tuffs and Rome

Rome occupies a striking, unique place within the rugged, mountainous Italian Peninsula, sitting as it does on a fertile plain crossed by the Tiber and Aniene rivers. When the very young volcanoes of Latium erupted large volumes of volcanic ash and pumice as pyroclastic flows that coursed down ravines and valleys, they filled valleys and actually changed the courses of the Tiber and Aniene rivers. The tuffs were easily eroded, yielding a landscape that is subdued, well-watered, and fertile. The rich volcanic soils provided grasses necessary for large herds and flocks of grazing animals, and the eroded tuff deposits became a countryside that was easily traversed.

The benefits of this terrain led to prosperity but also to trouble: wealthy settlements evoked the envy and cupidity of neighbors living in the hardscrabble limestone mountains near Rome, where defense was easy but making a living from the land was difficult. Soon, more secure sites were essential for the populations along the Tiber and on the tuff plateaus. The hills created by erosion now provided defensive locations as well as easily quarried building materials. What we term the "seven hills of Rome" are the remnants of the tuff plateau.

This geologic setting strongly influenced Roman architecture through the use of the local tuff deposits for construction stone. The volcanoes around Rome produced large volumes of ash and pumice (collectively, several hundred cubic kilometers; compare this to Mount St. Helens' output of one-half of a cubic kilometer in 1980), leaving deposits that are massive, mostly without fractures, and strong—yet soft enough to be easily cut, excavated, and shaped. It is not surprising that these rocks became a foundation of Roman construction practices. Roman tuff quarries have been found in many places, especially under the seven hills and along the Aniene River. Surface tuff quarries are especially well preserved in the Alban Hills town of Marino, where the tool marks are still evident from the 1-meter-thick (about 1-yard) blocks that were cut by ancient Roman quarrymen. Tuffs of several kinds were available to the Romans and were used in proportion to their accessibility and the ease of transport into the city.

Wherever tuff deposits exist throughout the world, from Italy to Mexico, ancient humans quickly learned to excavate spaces for living quarters, storage, and even burial of the dead. It has always been easier to excavate than to build. The Etruscans and then the Romans were no exception; they used natural cavities for shelter and excavated larger homes into the soft tuffs of Latium. Centuries of experience with tuff shelters and underground quarries may have taught the Romans that the most stable shape for such a cavity is that of a vault or arch—natural openings have this shape, and artificial ones acquire it through a process of collapse. Perhaps these observations led to the vault and arch as stable construction elements, freeing Roman architects from the severe limitations of the traditional column and lintel technique and ultimately leading to the architectural explosion that became a hallmark of Roman civilization.

In this surface quarry in tuff deposits in the town of Marino, in the Alban Hills, early quarrymen's tool marks are still visible along the wall. These quarries follow an ancient valley filled with tuff deposits. The Roman quarrymen cut blocks that were about 1 meter (a yard) thick to be transported to Rome.

Underground Roman stone quarries and catacombs were excavated, mostly in tuff deposits, in and around Rome. In the Tomb of the Scipios, located along the ancient Appian Way, the quality of Roman stonemasonry can be seen in the close fit between the tuff block and the underlying tuff deposit. University of Roma Tre engineers Prof. Carassiti and Dr. Brancalone examined the tomb to evaluate its stability and state of preservation.

Tuffs Used in Rome

Tufo pisolitico, the most common building stone used by people in Rome from the 6th to the 5th centuries B.C., was quarried in deposits left by eruptions of the Tuscolano-Artemisio volcano of the Alban Hills between 600,000 and 300,000 years ago. The deposits of fine-grained volcanic ash contain accretionary lapilli (spheres made up of concentric layers of fine ash so that they look like small onions or hailstones when cut open; also called *pisolites*), as well as trunks or the casts of trees knocked down by the rapidly moving pyroclastic flows. When exposed, these deposits have a dirty-brown appearance, which was caused by

Blocks of *Tufo pisolitico* were used to construct the Servian Walls, the first defensive walls around Rome, which were built in the 6th century B.C. You can see this wall remnant near the Piazza Albania, at the foot of the Aventine Hill.

water and steam that quickly altered small glass shards to clay and other finely crystalline minerals.

Pyroclastic flows emanating from the Tuscolano-Artemisio volcano reached the valleys of the Tiber and Aniene rivers, leaving sequences of 5- to 10-meter-thick deposits in what is now Rome. The *Tufo pisolitico* has been exposed in quarries throughout Rome, including on the slopes of the Palatine Hill. Early settlers excavated both shelters and tombs in these tuff deposits and quarried the tuff for building stone.

When improved transportation was available to the early Romans, their tried-and-true *Tufo pisolitico* was abandoned. After victories in the war with the Gauls and the conquest of Veii in 396 B.C., Rome acquired new territory north of the city, where the *Tufo Giallo della Via Tiburtina* is exposed. The Romans found that its strength made the *Tufo Giallo* a more effective building stone, and it was already being used for that purpose by the Etruscans. Quarries near the Tiber River made it possible to transport the tuff by barge into the city.

The *Tufo Giallo* was deposited by a large eruption of the Sacrofano volcano in the Sabatini volcanic field located about 30 kilometers north of Rome. This eruption occurred about 500,000 years ago and, like many other Latium events, was particularly energetic because it was hydrovolcanic. The tuffs were deposited by at least seven pyroclastic flows, which collectively covered an area of about 400 square kilometers and had a total volume of 8 cubic kilometers (2 cubic miles).

The *Tufo Giallo* was used throughout the Roman period; one good example is the altar of the Temple of the Deified Julius Caesar in the Roman Forum. The earliest known use of the stone was to restore the Servian Walls in 396 B.C. after they were damaged by the Gallic invasion. Where now visible, the restored walls are constructed of blocks 59 centimeters (23 inches) high, placed with the long dimensions alternately horizontal and vertical; some sections of the wall are as large as 10 meters (33 feet) high and 4 meters (13 feet) thick. Later restoration of the Servian walls occurred in many places, as we can see from imperfect junctions between *Tufo Giallo* blocks. The total length of the new (and restored) wall was about 11 kilometers (6.8 miles), enveloping an area of 426 hectares (about 1,050 acres); this enclosure made Rome the largest city on the Italian Peninsula.

Architectural guidebooks often refer to *peperino* tuff, which was quarried from two similar fine-grained deposits, the *lapis Gabinus* and *lapis Albanus.* The *lapis Gabinus* was erupted from the Gabii or Castiglione craters, and the *lapis Albanus* from the Albano crater in the Alban Hills volcanic field (where quarries are still active today). These tuffs are easy to cut yet provide relatively strong blocks. Both the *lapis Gabinus* and *lapis Albanus* peperino tuffs were commonly used throughout Imperial Roman times after a good transportation system had been established along the Via Prenestina and Via Appia.

Romans could now choose building stones for their strength and appearance, no matter how far the quarries were from the city. Today we can see numerous examples of these various tuffs from many periods

Tuff blocks in the foundation of the Temple of Antoninus and Faustina, in the Roman Forum, are partly *lapis Albanus*, one of several types of tuff favored by Imperial Roman builders.

of Roman history. The *lapis Gabinus* and *lapis Albanus* are well displayed in the Imperial fora. In the most ancient of these tuff constructions, the Temple of Antoninus and Faustina was constructed partly with *lapis Albanus.* In a later time, the Forum of Augustus was constructed largely with *lapis Gabinus* and *Lionato* tuff from the Alban Hills. *Lapis Gabinus* was also used for the foundation and walls of the Tabularium, which forms the base of Michelangelo's Palazzo Senatorio. The *Lionato* tuff was used for the back wall of the Forum of Augustus and for the substructure of the Temple of Mars Ultor, which abuts it. Contemporaneous use of both *lapis Gabinus* and *Lionato* tuff in the same period suggests that the two rock types were obtained from quarries that were not far apart. In fact, the *lapis Gabinus* quarries were located along the border of Castiglione crater close to the *Lionato* tuff quarries that were sited along the Via Tiburtina.

Monuments in the central area of Rome that include *peperino dei Colli Albani*

- Tullian Prison (3rd century B.C.)—lower rooms (possibly *Tufo lionato*?)
- Temple of Magna Mater (Temple of Cybele) (3rd century B.C.)
- Paving of the Forum (100–80 B.C.), close to Lacus Curtius and the Comitium
- Sanctuary of Sant'Omobono (100–80 B.C.)—foundations and external blocks
- Forum Holitorium (90 B.C.)—Ionic temple, both tuff and travertine; wall of the cella
- Tabularium (78 B.C.)—fluted half columns
- Forum Holitorium (80–50 B.C.)—Doric temple and foundation
- Temple of Saturn (42 B.C.)—podium, faced with travertine
- Forum of Augustus (ca. 25–2 B.C.)—back wall
- Temple of Mars Ultor (2 B.C.)—substructure
- Temple of Castor (131 B.C.)—Augustan restoration
- So-called Temple of Portunus (late 4th–1st centuries B.C.)—with other tuffs and travertine
- Temple of Antoninus and Faustina (A.D. 141)—external walls

(*From Bianchetti et al., 1994*)

The *Lionato* stone used during Roman times probably came from near Settecamini and Salone, where the local name is *Tufo dell'Aniene* (named for an area near the intersection of the modern Via Tiburtina and the Gran Raccordo beltway). Other quarries in the same deposit were closer to the city, at the foot of the Monteverde Hill, where the stone was called—yes, you guessed it—*Tufo di Monteverde.* The Monteverde quarries were used only for a limited time, however, because of the poor stone quality there and the constant danger of collapsing pit walls.

Sperone, a deposit of welded scoria (known as volcanic cinders in the United States), was formed during lava fountaining along linear vents that crossed the central Alban Hills; this deposit underlies the northern border of the Tuscolano-Artemisio volcano. The semimolten scoria fragments were welded when they accumulated around the crater. The massive reddish gray rock is easily cut and transported. This rock was used relatively little, however, probably because of limited access to and transportation from the quarries; nevertheless, we find it as the structural base of the Colosseum, the most important modern symbol of Imperial Rome. Although no Roman-age *sperone* quarries have been found, it is likely that they were located near the Alban Hills towns of Grottaferrata and Frascati, where they would have been close to roads leading into Rome. There are outcrops of *sperone* near the site of ancient Tusculum, where there is a well-preserved Roman village with a small theater, all built with *sperone.*

Preservation of the Tuff in Roman Monuments

In 1990, archeologists and experts on monument conservation came together in a rather distant location—Easter Island, Chile—to consider how tuffs were used for construction by ancient civilizations and how to preserve the products. Italian experts on stone preservation shared their experiences with those facing similar problems at sites from Mexico to Indonesia. Appropriately, the meeting was sponsored by the International Centre for the Study of the Preservation and Restoration of Cultural Property, which is located in Rome on the right bank of the Tiber. At the time Roman engineers quarried and transported these tuff blocks, wouldn't they have been astounded to know that their work would be the subject of an intellectual exercise in a new Western world they didn't know existed?

Although certainly adequate for their purpose and actually having stood the test of time, the tuffs used by Romans for construction are beginning to show signs of wear and tear. After their excavation, tuff blocks within the Roman city are deteriorating, partly because of their

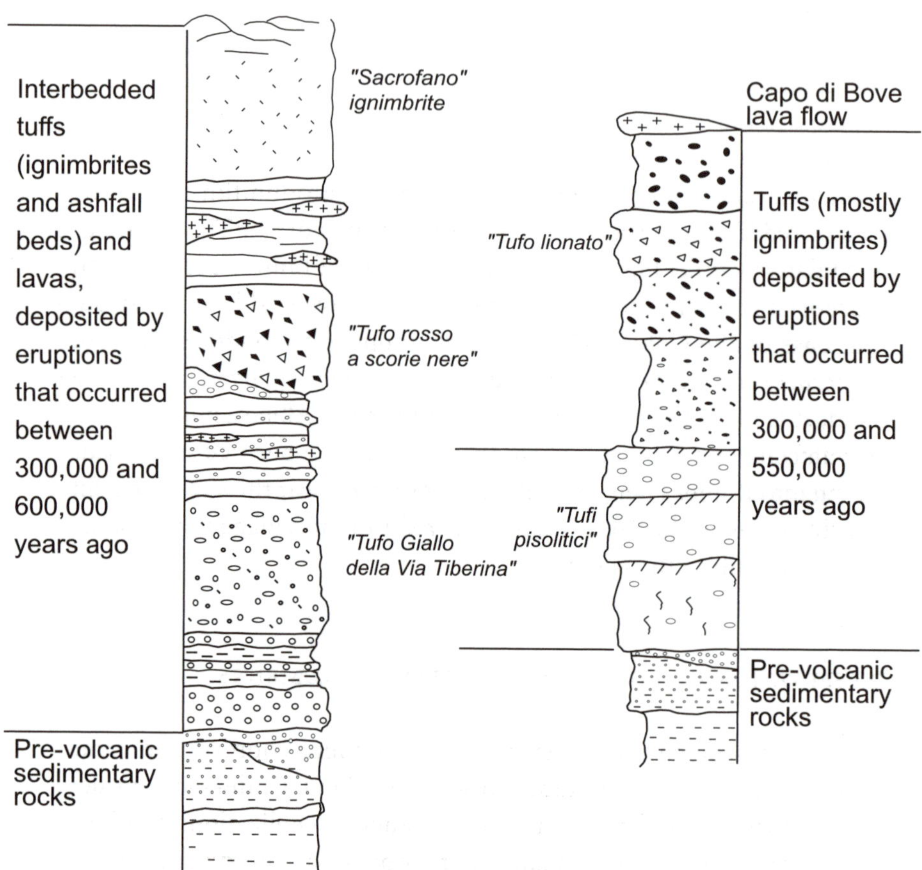

Volcanic rocks (mostly ignimbrites) for construction were brought to Rome from the Alban Hills and Sabatini volcanic fields, whose tuff deposits meet at the Tiber and overlap in places. The types of tuff highlighted in this sketch were (and still are) quarried for both building stones and feedstock for pozzolan concrete.

ability to absorb water and the heterogeneity of many of the blocks. Many examples are visible in the Forum below the Palatine Hill. One of the most interesting is the foundation of the Temple of Antonius and Faustina, where you can see the original bedding left by the pyroclastic flow that formed the tuff—great stuff for a volcanologist trying to un-

You can see tuff outcrops along the Via della Consolazione, along the southern base of the Capitoline Hill. The hill is composed of several tuffs (pyroclastic flow deposits) from the Alban Hills volcanic field.

ravel past volcanic activity, but of serious concern to a construction engineer or preservationist. The coarser grained layers in these tuff blocks have been eroded by wind and water, leaving the finer grained layers to stand out as ridges close to the original quarried surface. As long as these blocks are kept moist, they appear to retain their shape (one reason many were used to line cisterns and sewers, which are so well preserved that they could be used today if needed). Conservationists' studies at Italian universities and elsewhere are examining possible ways to preserve the crumbling tuff blocks. The work to date, using chemical approaches, has taken place in the laboratory; field tests have been limited. The problem facing these researchers is how the chemical treatments would stand the test of time. Would these blocks of tuff last another 2,000 years if treated, or would they best be left alone?

• • •

On the Palatine Hill, the great palaces of Septimus Severus, the stadium, and the Domus Augustana were all built of brick on older stone buildings that were constructed of tuff blocks. Although tuffs form much of the foundation of Rome, bricks and concrete are some of the most visible construction materials used throughout its history.

By this time, you may be looking for some peace and tranquillity. Cross the Circus Maximus and walk to the public rose gardens located on the northeastern slope of the Aventine Hill. This is your entrance to the Aventine, one of the seven hills that is now characterized by quiet neighborhoods, churches, and parks.

CHAPTER **4**

The Aventine (Aventino) Hill

We descended a good depth in the bowells of the Earth,
a strange and fearefull passage for divers miles.
—John Evelyn (1620–1706), *Diary*

The Aventine Hill is larger (0.4 square kilometers, or 96 acres) and somewhat more geologically complex than the Palatine. In a few respects, however, the Aventine mirrors the Palatine's form and is separated from it only by the small valley that is the Circus Maximus. The southernmost of the seven hills and closest to the Tiber, in Roman times the Aventine was dissimilar to the Palatine in that it was a residential area for middle-class citizens. Today it is one of Rome's most elegant neighborhoods.

Like the Palatine Hill, the Aventine is easily recognizable as a plateau from all sides. The most interesting and accessible approach is from the Circus Maximus to the Piazzale Ugo de Malfa, where streets radiate to the west across the Aventine. The Aventine is an excellent point from which to view the Circus Maximus, the Palatine Hill, and segments of the historical center of Rome. On clear days, you'll see the volcanic field of the Alban Hills, the source of most of the pyroclastic flows that make up the seven hills of Rome.

One of the largest of the seven hills, the Aventine is a little different than the rest. Much of the northern slope, adjacent to the Circus Maximus, consists of a sandy claystone in the form of a low hill that was later buried by pyroclastic flows (tuffs) from the Alban Hills volcanoes. The main part of the Aventine plateau is underlain by the tuffs, which filled and leveled the older terrain. Romans often quarried tuff underground for building stone, and the Palatine Hill is permeated with ancient Roman quarries. You could justifiably ask, "When it was readily available above ground, why did the Romans use underground quarries?" Economics. Because surface land was so valuable and because open quarries were heavily taxed. To see tuffs used for a defensive wall, examine remnants of the Servian Wall along the southern edge of the Aventine Hill at the Piazza Albania.

Ancient City Walls

Cities need protection during troubled times, and eventually defensive walls were constructed to encircle Rome. King Servius Tullius reputedly erected the first Roman walls during the 6th century B.C., and remnants are still visible at the base of the Aventine Hill near the Piazza Albania. The Servian Walls were constructed of *Tufo pisolitico* blocks quarried within the city. Some of the quarries for these blocks are still visible in archeological excavations below the Termini Station and at the eastern edge of the Aventine Hill. Defensive walls followed the Pomerium, a sacred line that marked the perimeter of a Roman settlement. The Pomerium was marked by boundary stones and blessed with sacred ceremonies; there could be no construction, cultivation, or burial along this line. However, cities grow, and the lines evolved. Originally, this line encompassed only parts of the seven hills, but it obviously expanded with the growth of Rome.

Little is left now of the Servian Wall built with the relatively soft *Tufo pisolitico*, but its construction marked the beginning of 2,300 years during which Rome constructed, destroyed, and reconstructed its defensive walls. In some walls, more compact, stronger stones from a variety of tuff deposits were stacked in horizontal rows (*opera quadrata*). After being arranged in rows, the blocks were bound with "double-T" metal clamps. Quarrymen marked the blocks with Greek letters to keep track of their work when it came time to collect payment for their services. You can see examples at Termini Station and on the Aventine Hill near the Piazza Albania.

Underground Rome is vitally interesting to diverse groups: officials of the City and Province of Rome, who worry about the risk of streets or buildings collapsing into an ancient tunnel; archeologists, who are always excited by a newly discovered necropolis or catacomb; business opportunists, who view the underground as a space for commercial storage or growing mushrooms (an important ingredient in Roman cooking); and a new breed of urban speleologists, who simply enjoy the challenge of underground exploration.

• • •

The pleasant Via San Giosafat, on the Aventine Hill, is underlain by ancient quarries. These Aventine quarries are widespread and extend under the Piazza Albania. Italian geologists Walter Santoro and Vittorio Federici have been investigating an ancient quarry complex under the southeastern part of the Aventine by drilling exploration boreholes and examining the samples they collect to determine rock types and the properties that define rock strength. First, the geologists lowered cameras through small boreholes to view the tunnels. After a preliminary evaluation, a few wider bores (of about 1-meter diameter) were drilled, and the geologists themselves were lowered into the tunnels to survey and evaluate the stability of the tuff quarries. This approach involves careful geologic mapping, precise engineering, and a bit of "Indiana Jones" (without the snakes).

The quiet, beautiful Aventine Hill neighborhood provides no indications of a serious hazard—possible collapses into ancient underground quarries, which are a problem throughout Rome. As old as many of the streets and buildings of modern Rome may be, they are still the youngest of many historical layers that make up this city, and they mask the labyrinth of tunnels that lies underneath. Despite valiant efforts by geologists, engineers, archeologists, and utility workers, occasionally buildings and streets collapse into underlying tunnels.

No one knows precisely the magnitude of Rome's underground real estate. In some respects, trying to determine its size is similar to guessing the extent of an ant colony on the basis of the surface hole and mound. Why is Rome underlain by a labyrinth? Over the last 2,800 years, tunnels have been excavated for both economic and funerary purposes—chiefly for construction materials and places to inter generations of the dead. In addition, there were underground springhouses, cisterns, drains, theaters, houses, villas, churches, and, recently, utility tunnels. All the tunnels are man-made; none are natural. Many underground cavities are mentioned in historical records, but just as many were buried and forgotten after the destruction that accompanied earthquakes, fires, and invasions.

Rome is underlain by a labyrinth of tunnels that were cut for the purpose of quarrying the stone or for catacombs, shrines, and chapels. Today, these spaces are attracting interest from widely diverse groups: archeologists, safety experts, business opportunists, tourists, and speleologists. In the Tomb of the Scipios, spaces were carved in tuff deposits along the Appian Way. Tool marks are still visible on the ceiling of this tomb, which was active from the 3rd century B.C. to the 1st century A.D. The tomb was discovered 300 years ago by an Appian Way cantina owner who was digging a new basement.

One of the most famous underground structures in Rome is now open for tours. When Nero's "Golden House" (Domus Aurea) was built after the great fire of A.D. 64, the buildings covered 200 acres of the Esquiline Hill. No expense was spared in creating this sumptuous Imperial home. After Nero's suicide in A.D. 68, the Domus Aurea was forgotten (*damnatio memoriae*) to erase the memory of the unpopular emperor. The palace was progressively covered by portions of the Colosseum and by the Baths of Titus and Trajan. Only some subterranean rooms of the Imperial apartments have survived, along with well-preserved frescoes, and are accessible along the Via Labicana below the Oppian Hill.

Contrary to the ideas put forth in popular movies, tunneling in ancient times had little to do with intrigue or religious freedom; it occurred simply because much of Republican and Imperial Rome was underlain by thick tuff deposits. The relatively soft tuffs were easy to excavate and, as discussed earlier in this book, provided reasonably good (and cheap) building stone. Underground quarries may have several levels and are usually within the first 4 to 15 meters (13 to 50 feet) below the ground surface. To safely cut and remove rock, Roman quarrymen used a "room and pillar" technique—something like an exaggerated version of an underground parking garage, except that the quarries' stone pillars are much thicker than the concrete pillars of a parking garage. Rooms are between 2 to 3 meters (7 to 10 feet) wide and 2 to 5 meters (7 to 16 feet) high, and the pillars are 5 to 25 meters (16 to 80 feet) on an edge. With the huge demand for construction stone in Rome, the tunnel networks eventually grew to underlie all the tuff plateaus, including the seven hills. Although the quarries were relatively safe when excavated, some of the stone pillars are beginning to degrade and fail, especially after centuries of constant vibration and irregular weight distribution of the evolving city overhead.

The Catacombs, multilevel tunnel networks excavated to inter the dead, expanded rapidly when Christian and Jewish communities of Rome began to choose burial rather than cremation, which generated a land shortage for above-ground tombs and cemeteries. Another factor might have been a shortage of fuel wood that increased the expense of cremation. Catacomb internment was an efficient use of space, and it allowed families to place the deceased faithful near a sacred place or the remains of a saint. Eventually, the Catacombs became tourist attractions. For example, in the Jubilee Year of 1450, the catacombs of San Sebastiano attracted tens of thousands of pilgrims.

You can see how the geologic setting made it easy to excavate catacombs if you follow the Appian Way to the catacombs of San Sebastiano and San Callisto (Callixtus), the oldest official Christian cemetery in Rome. Walking from the city, you pass under the city wall at Porta San Sebastiano and up a gently rising plateau, formed when pyroclastic flows erupted from the volcanoes of the Alban Hills. Within this tuff, the catacombs of San Callisto were established by the Deacon Callixtus in A.D. 199 and then expanded when he was pope (A.D. 217–22). These catacombs once held the remains of the nine popes who reigned between A.D. 230 and 283; all remains were later moved to the Vatican.

The catacombs of San Callisto are on four levels, the deepest of which is 20 meters (65 feet) below the surface, and are composed of 20 kilometers (12 miles) of narrow galleries along which the dead were in-

Rome's underground cavities are historically important, but they present a potential hazard for the modern city, as shown in this diagram.

terred in *loculi* (niches) above the gallery floor. After an internment, the loculus was sealed with stone and plaster. The stability of the narrow galleries is evident; there are few signs of collapse during the last several thousand years. Some of the catacombs were discovered by accident (the catacombs of San Callisto were encountered in 1578 by a workman digging in a vineyard), and it's possible more will be found in the future (one hopes *not* as the result of the collapse of overlying apartment buildings). Tunnel complexes also underlie the Via delle Sette Chiese and the Via dell'Arco Travertino, streets that pass through modern suburbs along the city's southeastern edge. Complex tunnel networks beneath the many densely packed apartment buildings raise questions about the safety of the area.

When engineers Federico Pagliacci and Maurizio Conti were contracted to evaluate an ancient tunnel system underlying the Santa Beatrice School on the Aventine, they employed some modern techniques. You can imagine the scene. Watching workers drill into a cavity at perhaps a few meters an hour, more or less, you get into a somnambulant rhythm—the penetration rate and pitch of the spinning pipe create a comfortable feeling about extracting a good core. It is a time for

It's easy to recognize the potential dangers posed by unstable tunnel networks in this photo of a Roman street that has collapsed into an underground cavity.

daydreaming. Suddenly, the drill string drops abruptly after entering a cavity, disturbing this reverie, shaking the drill rig sharply, and surprising the driller, judging by the expletives he emits. The process is repeated again and again in a series of small-diameter boreholes that will identify the basic outline of the underground tunnels. Now, workmen insert a small, high-resolution television camera and a rod with meter marks (for scale) into adjacent boreholes. By scanning 360 degrees with the camera, they produce a group of images they can use to map the tunnels without actually entering them. With a general idea of the shape and size of the tunnels, the engineers develop a plan for stabilizing the cavities. Workers insert stiff but expandable plastic tubing into the holes and pump concrete through the tubing to form supporting columns that will, engineers hope, reinforce these ancient tunnels.

Archeologists in Rome view the underground more as a part of the cultural heritage than as a geotechnical hazard. One of the goals of the City of Rome is to identify and preserve tunnel systems, especially if they involve underground cemeteries or temples. Other important features of Roman heritage are the tunnels carved for ancient aqueducts and water distribution systems or for drainage systems.

In addition to the professionals evaluating underground Rome, a determined group of volunteers make up the first group of urban speleologists (known in North America as "spelunkers"). This group is adding to the wealth of information about Roman history by creating Web pages and handbooks on subterranean Rome, which sun-weary tourists can use as guides or for a virtual tour to Rome's more famous subsurface attractions.

The subsurface of Rome needs systematic mapping to identify and evaluate all ancient quarries, catacombs, and tunnels used for aqueducts. Depending on your interest—engineering geology or entrepreneurial possibilities, for example—the ancient underground quarries can be a fascinating historical resource or a menace. Rome's underground should be evaluated for the potential collapse hazard and as possible space for commercial development. Despite 3,000 years of history, there is still challenge in the exploration of Rome.

• • •

While on the Aventine Hill, visit the early Christian Basilica of Santa Sabina, which overlooks the Tiber River. The park next to the basilica provides excellent views across the Tiber's floodplain into Trastevere and the Vatican. From this vantage point, it is easy to see why the Tiber has been so important to Rome—both as the chief means of transport and as the source of catastrophic floods.

CHAPTER 5

The Tiber Floodplain, Commerce, and Tragedy

> The City of Rome is nearly all inundated by water. Along the riverbanks, since 1846, there have been no similar cases where water has risen over a meter above the average level of the river. The Via della Tipografia, where our newspaper is printed, is also under water. The water reached a meter. Piazza Colonna, toward 11 A.M., was under water. At present, flooded are: La Lungara, Piazza della Rotonda, Piazza Pia, Piazza Navona, Piazza Giudia, Piazza Montanara, Il Ghetto, namely, Via Fiumara, Piazza S. Andrea della Valle, Piazza S. Eustachio, Piazza and Via del'Orso, Piazza Campo de' Fiori, and the street of Via Tor di Nona, the Ponte S. Angelo at the Montebianco end, along the Piazza dell' Orso, S. Lorenzo in Città.
> *Il Tribuno*, December 28, 1870

Without an excellent map and a good sense of direction, you can become thoroughly disoriented as you explore the area of Rome that is on the Tiber's floodplain. One solution is to stay on the busy streets that follow the riverbank (the Lungotevere, "along the Tiber"). The scenery is superb, and you can spend days studying the bridges, but you must eventually enter the districts, such as Trastevere or the Campo de' Fiori, that sprang up along the margins of one of the world's best-known rivers. In search of famous sites in this area, nearly every visitor to the Eternal City ends up at the Pantheon. We begin our visit to the floodplain neighborhoods at Santa Maria Sopra Minerva, which is close to the Pantheon.

• • •

One of the few examples of Gothic architecture in Rome, the Church of Santa Maria Sopra Minerva is in the heart of the historic district.

This splendid old print shows the Piazza Navona during the flood of 1870.

Dating from the 13th century, this beautiful church was built on ancient Roman ruins and possibly on a temple of Minerva. It contains the ornate tombs of the 16th-century Medici popes Leo X and Clement VIII, the painter Fra Angelico, and Saint Catherine of Siena.

In addition to its historical and artistic treasures, Santa Maria Sopra Minerva displays a record of flooding and highlights the uneasy relationship that has long existed between the city and the Tiber River. Stone tablets attached to the church facade by the Dominican order

Plaques on the facade of the Church of Santa Maria Sopra Minerva mark the water depths during major floods of the Tiber River. The highest plaque records a water depth of 3.95 meters (13 feet) in the Piazza di Minerva for the flood of December 24, 1598—an event that devastated central Rome.

chart water depths during past floods. The worst inundation occurred on December 24, 1598, when floodwaters were 3.95 meters (13 feet) deep in the Piazza di Minerva. Try to imagine your favorite tourist destinations surrounded by water, including the Pantheon, Piazza del Popolo, Piazza Navona, and Castel Sant'Angelo.

Stone tablets located throughout Rome on both sides of the Tiber record the highest levels reached by floods. A systematic flood record remains in the Piazza del Porto Ripetta, on the left bank of the Tiber near Ponte Cavour, a minor landing place for boats and barges when the river was still navigable. Lest the citizens of Rome forget about the flood threat, the architect Alessandro Specchi constructed the Ripetta

The Ripetta column is located in the Piazza del Porto di Ripetta, a former river port on the left bank of the Tiber, near the Ponte Cavour. The column was constructed by the architect Alessandro Specchi in 1704 to mark water levels during floods of 1495, 1537, and 1660 and is annotated with the names of popes during those flood years.

column in 1704 to mark water levels during the floods of 1495, 1537, 1660, and 1805. Much of what we know about these Tiberian floods is based on a thorough and patient integration of historical records kept by the National Hydrographic Service and the Hydrographic Office of Rome. Fifteen floods were described by Roman writers, including Livy, who wrote of a flood in 192 B.C.: "The river with much force and violence, much greater than in the previous year, burst through the city, knocking down over two bridges and many buildings—above all in the Porta Flumentana" (from Bencivenga et al. 1995).

Damage during Imperial Roman times was less devastating because of competent city planners, who placed facilities such as theaters and athletic training centers on the floodplain and located most residences above the rogue waters. During medieval times and later, uncontrolled growth extended the city's neighborhoods across the floodplain to the Tiber's banks, and eventually much of the population was exposed to flooding. Records from medieval times are spotty and include emotional observations such as "serpents and dragons in the river." A more objective and realistic—albeit less entertaining—description of the flood of 1277 is inscribed on a stone tablet located under the L'Arco dei Bianchi: "Arrived here the turbid Tiber and withdrew quickly in the year of Our Lord 1277 from the second to the seventh day of the month of November, when it withdrew" (from Bencivenga et al. 1995).

The effects of floods highlighted by the stone tablets of Santa Maria Sopra Minerva are clearly and succinctly presented in writings of their times:

> *October 8, 1530.* "There was already at sunrise on Saturday morning, the eighth of October, when the Tiber moved out of its usual bed, beginning 'mountains' of water, much to the surprise of everyone—Thus began the wretched river's flooding of the city before midday and reaching all the sewers, the cellars, and the lowest places, a little after was seen the water overcoming the houses like a betrayal, and hidden power with torrents beginning to seize the steps and undercut all of the roads with great fury that seemed to chisel away at the city's foundations." (L. Gomez, 1531, from Bencivenga et al. 1995)

> *September 15, 1557.* "Toward midnight, the Tiber rose and flooded a large part of the City. This sudden, unexpected catastrophe did not allow anyone

time to save their possessions. The vineyards near the Castel Sant'Angelo were eroded away by the violent currents and the inhabitants took refuge on their roofs. On the left side in the ditch, water reached the heights of 1530. In the Plaza of St. Peter's you could go by boat. After 24 hours the water began, little by little, to recede and the poor then evaluated the damage. There was total destruction of the Ponte di S. Maria and the new moles along the Tiber. There was damage to the Ponte Fabricio, the passage from Castel Sant'Angelo to the Vatican and the new fortification of the city. There was near collapse of the Convent of S. Bartolomeo on the Tiber Island and in some cases, the palazzos." (L. V. Pastor, *Storia di Papa*, vol. 4, 1944, from Bencivenga et al. 1995)

December 23, 1598. "The evening of the 23rd, the month of December past, the Tiber began to rise out of its bed, no longer below the city, rising continuously till 10 o'clock the following night, putting the city under water, outside the seven mountains and the summits of some places higher in the middle of the city, surpassing the riverbanks and signs of flooding in ancient and modern times, particularly, the plain, more than that which came in the time of Pope Clemente VIIth, in a note from year 1530, was the ruin of the Bridge of Santa Maria, the two arches outside parts of the Ponte Molle, and that of Sant'Angelo, even though it remained, the fury of the water struck the shacks and workshops, this directed at the Castle, which fell in places and was ruined—Submerging forty valuable places that were in Torre di Nona and the countryside—drowned many persons and animals, large and small. This horrendous spectacle was finished by 4 P.M. and in this hour, which was the birth of our Lord Jesus Christ, began to go down, and by Christmas Day, was down three levels. It is said that the sudden flood drowned in this city and surrounding areas 1,400 people." (Anonymous, 1599, from Bencivenga et al. 1995)

Even as recently as the 19th and the 20th centuries, the floods have caused devastation. There were twenty-eight floods—the highest in 1948, with a flood crest of 18 meters (60 feet). The largest stream flow (with 3,300 cubic meters, or nearly 900,000 gallons, per second) was measured during the flood of 1900, which caused severe damage in the city. However, a more complete warning system had been established after the turn of the century, saving lives along the Tiber's banks.

Why So Many Floods on the Tiber and Its Tributaries?

As you walk along paths and streets near the Tiber, notice the walls erected to keep floodwaters from damaging the low-lying parts of the city. Looking at the normally placid river, few people think about its origins. The Tiber River basin is one of the largest on the narrow Italian Peninsula. Because of the area's geologic structure, much of the Tiber's 403-kilometer (250-mile) length runs not perpendicular to the peninsula but parallel to the Apennines across Tuscany, Umbria, and Lazio before entering the sea at Ostia. Much of the Tiber's drainage is located north of Rome, beginning at an elevation of 1,268 meters (4,160 feet) at Monte Fumaiolo and covering an area of 17,156 square kilometers (6,623 square miles) before it reaches the sea near Rome's Fiumicino (Leonardo da Vinci) Airport. In this drainage area, forty-two northern and eastern tributaries feed the Tiber.

Many of the tributary streams south of what is now Rome were cut off when volcanic eruptions in the Alban Hills formed an enormous "dam" of deposits. After the eruptions formed the massif of the Alban Hills, many relatively small streams from their northern and northwestern slopes added their runoff to the Aniene and Tiber rivers. The Alban Hills streams that flowed into Rome have since been diverted or covered during the city's growth. Several of the southern streams, now buried under Rome's suburbs, eroded the tuff plateaus, forming the southern and eastern topographic boundaries of the ancient city; La Marranella flowed northwest, then north, emptying into the Aniene River; it now underlies railroad lines that go north from the city. The Marrana della Caffarella flowed northwest, then west along the southern city wall, under what is now the Central Market, and into the Tiber north of the Basilica of San Paolo fuori le Mura. One of the Alban Hills streams was the Tor Marancia River, which has been diverted to flow under the Via Ardeatina, nearly parallel to the Appian Way. Streams immediately north of the city were less affected by urban growth and are still visible.

Anyone who has lived in central Italy for more than a few months realizes that the weather is tempestuous and difficult to forecast. The Italian Peninsula has all the ingredients necessary to produce flooding: it is flanked by two seas and has a high, mountainous "backbone"—

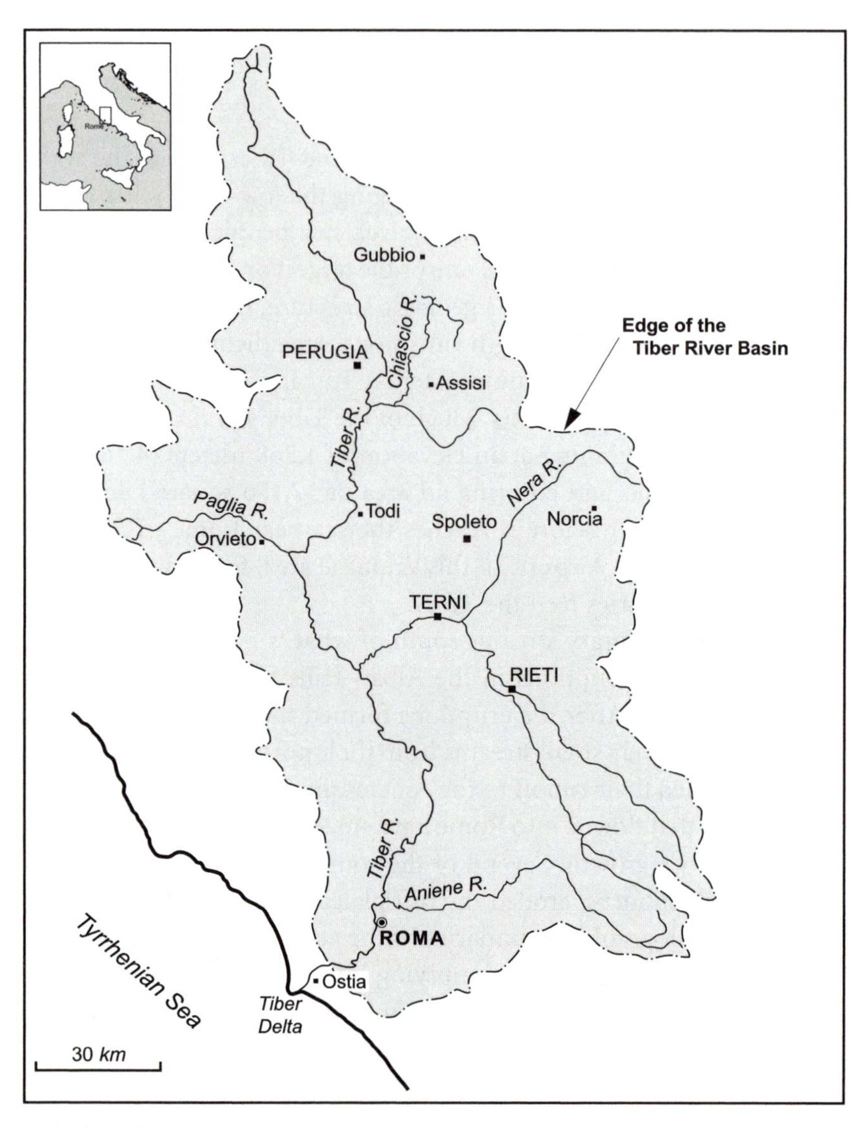

The Tiber River basin is a large north-south network that drains the western slopes of the central Apennines and the volcanic fields between Rome and Tuscany. (Adapted from Bencivenga et al. 1995)

ideal conditions for intense rainfall focused in relatively small areas. Storm clouds laden with moisture come ashore and rise over the central Apennines; the rising, moist air is cooled and begins to condense, producing local rain showers. If the ground is already saturated before a heavy storm, the rainfall flows quickly into rivers rather than being absorbed.

For three days in December 1937, a heavy storm passed over the Tiber basin; total rainfall exceeded 200 millimeters (8 inches) in the mountains north and east of Rome. The subsequent flooding was not the largest that Rome has seen but was important because it was one of the first for which systematic rainfall data were collected. Beginning on December 16, the Tiber reached a level of 16.9 meters (55.5 feet) at the Ripetta column, flowing through the city at 2,800 cubic meters (740,000 gallons) per second. The flooding was recorded on a stone tablet located near the entrance to the Church of San Bartolomeo on the Tiber Island, where water was 80 centimeters (32 inches) above ground level.

Try to imagine it: 17,156 square kilometers (6,623 square miles, an area nearly equal to that of New Jersey) is a lot of territory draining into a single river that runs through Rome. To understand this runoff system, city planners had to integrate many data sets: first, the meteorologic conditions, especially the movement of storms and masses of moist air; second, the topography, which controls the orographic effect (air that is cooled as it rises over the mountains and condensation of water that leads to rain); third, the hydrologic character of geologic units and soils in the region to determine how much water could be absorbed and how much would run off; and last, but not least, a network to monitor rainfall and the stream flow of the Tiber and its many tributaries. Such a sophisticated hydrologic network, now maintained by the National Hydrographic Service, is used for estimating recharge of aquifers and water to be stored for hydroelectric facilities, as well as to determine if and where there will be flooding. Because so much of the runoff from this basin enters the Tiber and passes through Rome, the city has a considerable stake in the success of this monitoring network. Catastrophic floods could destroy not only many neighborhoods but also precious artwork and architectural treasures.

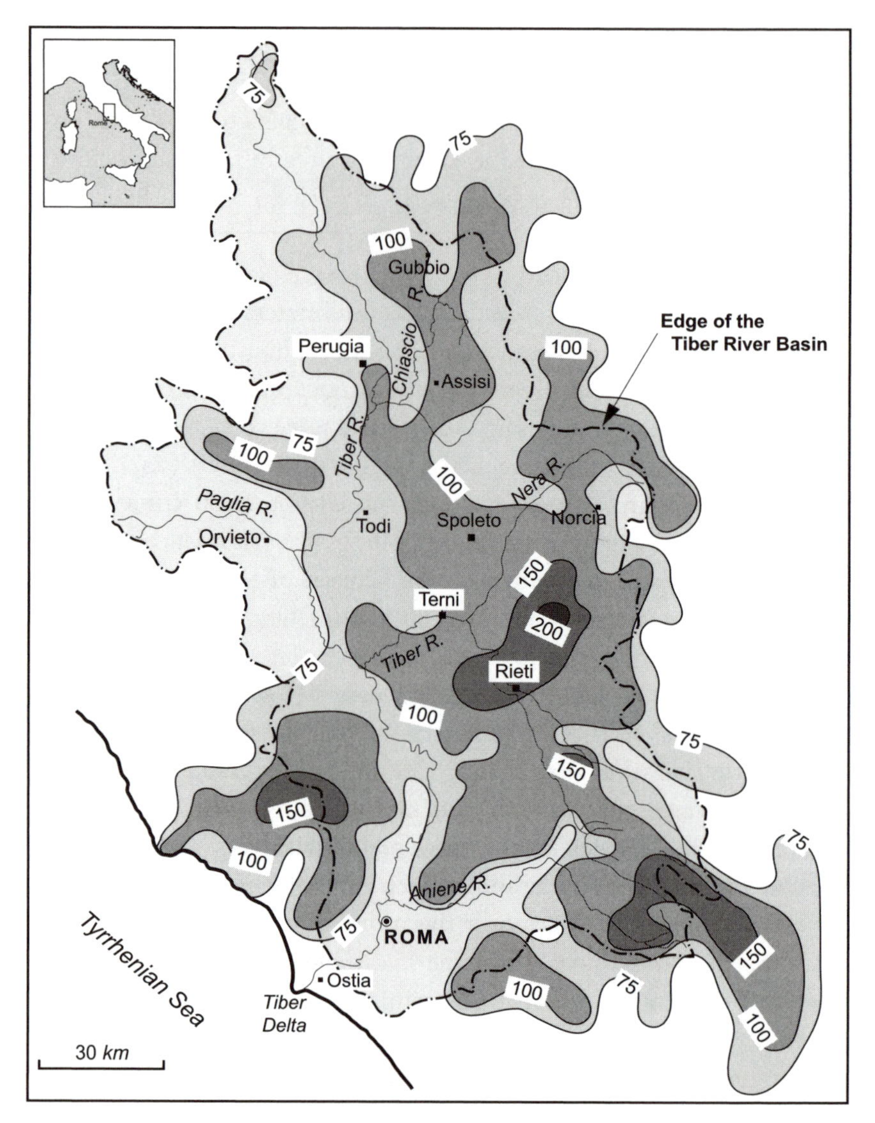

Experts have reconstructed the distribution of rainfall over the Tiber basin during the storms of December 16 and 17, 1937, which caused flooding along the Tiber and in Rome. The lines superimposed on the map of the basin show precipitation measured in millimeters. The most intense rainfall was near Rieti and in the mountains above Subiaco, where more than 200 millimeters (8 inches) fell in three days. (Adapted from Bencivenga et al. 1995)

The Pantheon and Early Utilization of Concrete

After your visit to Santa Maria Sopra Minerva, it would be impossible not to spend some time in the nearby Pantheon. This remarkable building has survived flooding, modifications, and long periods of neglect over several thousand years. Emperor Hadrian (A.D. 118–125) ordered the Pantheon built on the foundations of a temple that had been founded by Marcus Agrippa. In A.D. 609, Boniface IV consecrated the building as the Church of Santa Maria ad Martyres. In 735, the dome was roofed with lead sheets. More or less abandoned from 1305 to 1377, the sacred Pantheon was used as a poultry market.

What is the chief reason for the Pantheon's longevity? It was built with concrete. The building's dome is 43.3 meters (140 feet) high and 43.3 meters (140 feet) in diameter. The base of the dome, where it rests on load-bearing walls, is 7 meters (23 feet) thick. As it rises toward the opening in the crest, the dome gradually thins to 2 meters (6 feet). In addition, the concrete mix changes from cement with a rock aggregate at the base to cement with pumice (the very lightweight product of explosive volcanic eruptions) at the top. The engineers also used hollow cast-concrete coffers, which reduced the material's weight but not its strength. The real secret of the long-lived dome is that it is made of *pozzolana* concrete, which has resisted weathering, earthquakes, and poorly conceived modifications by later city leaders.

The Romans' discovery of concrete had a profound influence on Western civilization. Concrete was a stable mortar for bonding stone and brick masonry, and its durability allowed builders to construct cast-concrete structures that have survived both the climate and earthquakes for several thousand years. Roman engineers found that when they combined fine-grained, weathered volcanic ash (or crushed tuff) with lime cement, they could create pozzolan (*pozzolana*), a strong, water-resistant concrete. The Romans called the weathered ash deposits *pulvis Puteolanus* because their first source was Pozzuoli (Puteoli), a town on the Bay of Naples. Densely populated modern Pozzuoli is situated in the middle of the Phlegrean Fields, a young volcanic field where there is not only weathered, fine-grained tuff but also considerable volcanic

The Pantheon is one of the world's best-known and oldest concrete structures. Constructed of pozzolan concrete (a mixture of crushed, weathered volcanic tuff, calcined lime, and water), the dome has survived changes in power, politics, invasion, and cultures.

risk. Weathered tuffs similar to those that underlie Pozzuoli are also present below Rome, including the seven hills.

The Roman engineer Vitruvius specified a ratio of one part calcined lime to three parts of crushed *pozzolana* for underwater structures. After many centuries, the ratio for pozzolan concrete used today for marine structures remains much the same. Pozzolan cement can harden underwater, a factor that makes it useful for lining canals and cisterns, as well as for piers, seawalls, lighthouses, and breakwaters. Pozzolan is still used in and around Rome; the quarries are located in the tuff deposits of the Alban Hills, just beyond the Gran Raccordo Annulare, Rome's ring road.

Tiber Island

In the middle of Rome, a boat-shaped island divides the Tiber River. The 300-meter-long (984-foot) Tiber Island can be seen from the Ponte Garibaldi, the Ponte Palatino, or major streets that follow the river's banks. It can be reached by either of two small bridges: the Ponte Fabricio, on the eastern side, is a pedestrian-only bridge that offers respite from motor vehicles; the Ponte Cestio provides access from the west bank. Tiber Island has existed at least since 293 B.C., when a temple was constructed there to honor Aesculapius. The island served as a sacred site for centuries, providing a home for many churches, including the Church of San Bartolomeo, constructed in the 10th century A.D. Today more than half of the island is occupied by a modern hospital. The Tiber Island has retained more or less the same shape for 2,300 years. In Rodolfo Lanciani's reconstruction of ancient Rome, he shows another smaller island, which disappeared following the construction of walls along the river and infilling along the riverbanks (probably in the early 19th century, when the smaller island ceases to appear on maps).

So why is there an island here? This is the only island along the Tiber in the immediate vicinity of Rome, and it's located within a meander (a bend in the river) where the floodplain is narrow (675 meters, or 2,200 feet). The outside of the meander cuts into rocks of the Capitoline Hill on the east side and the Janiculum Hill on the west side of the Tiber.

A modern pozzolan quarry near Rome is located just south of the Gran Raccordo Annulare (GRA), Rome's ring road, in fine-grained pyroclastic flow deposits from the Alban Hills volcanic field.

Most river channels meander when there is equilibrium between water flow, deposition of sediment, and erosion. The river channel is usually deepest along the outside of a meander because centrifugal forces claw at the riverbank when water rounds the curve. Lower current velocities on the inside of the meander allow sand and gravel to accumulate and form a gravel bar or sandbar, which gradually enlarges to form an island.

The Tiber Island is located at a point where two or perhaps three tributaries once emptied into the Tiber. It's likely that the extra sediment load from the tributaries increased the size of the gravel bar

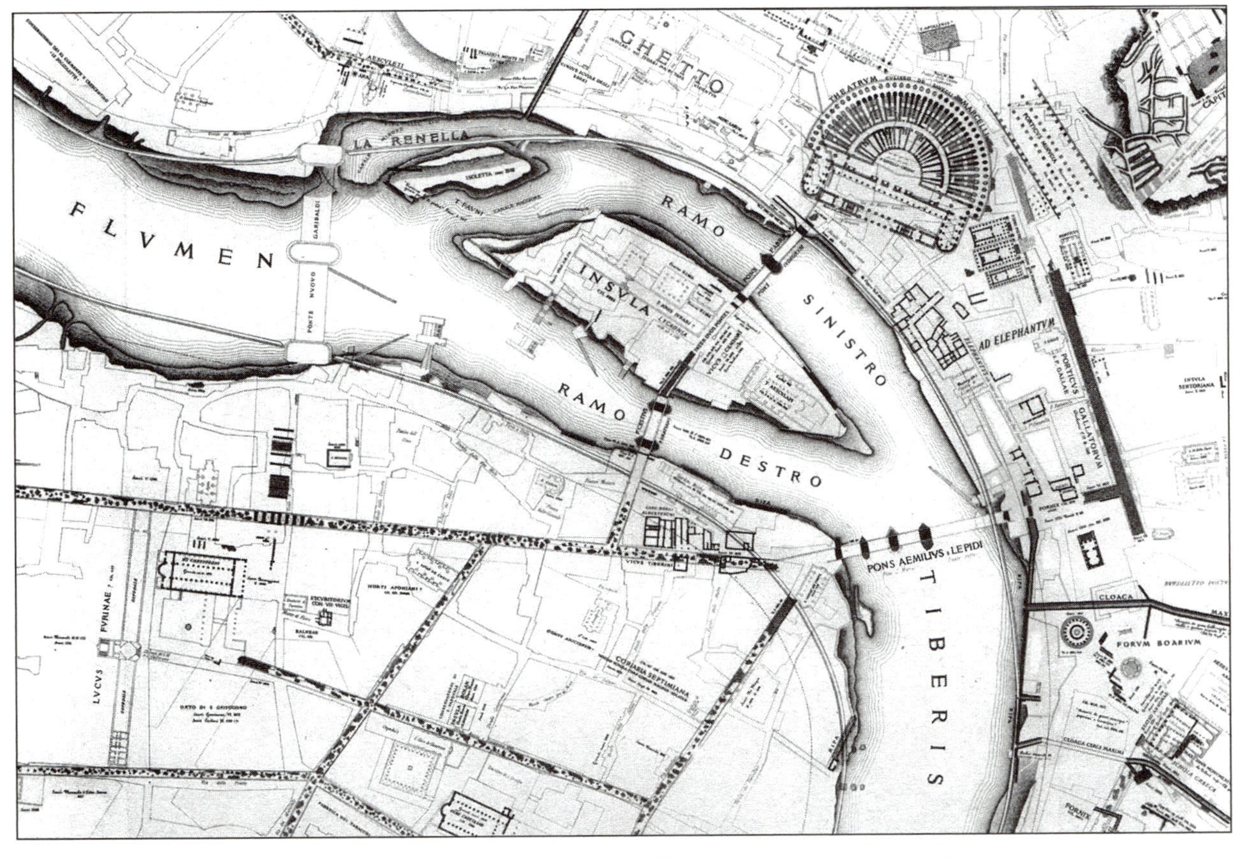

This drawing of the Tiber Island during the time of Imperial Rome (map by Lanciani) shows that the basic shape of the island is unchanged today. The small islet (*isoletta*, below the La Renella channel) was there at least until 1807. That small channel between the islet and the riverbank was later filled in and is now below the street that follows the riverbank.

In this later drawing of the Tiber Island in A.D. 1676 (from a map by G. B. Falda), the island is connected by the Roman bridge (Ponte di Santa Maria) immediately downstream. The bridge was finally destroyed during the flood of 1598, and the remnant is now named the Ponte Rotto.

to the point where it became an island that mimics the shape of the curving meander. The river then could have split during a flood, cutting an additional channel across the inside of the meander and creating the island we see today. Early Romans reinforced the island with rock, protecting the boat-shaped island's "prow"; eventually the entire island was clad with rock and concrete (you can see this best from the Ponte Garibaldi, at the foot of the Viale di Trastevere). This answer to your question is actually geologic speculation, but certainly it is as valid as any of the other proposed explanations, which include the possibility that the island was built of waste from grain ships or from an overloaded ship that sank in the river. The problem with these speculations is that the island was present before heavy merchant activity, and it appears even on a Roman map ca. A.D. 200 and in the earliest written descriptions.

Floods and Climate Change

Until the Ripetta hydrometer was installed in the Largo San Rocco in 1821, estimates of floodwater volume and extent of flooding in Rome were only qualitative, but they were good enough that we can speculate about a link between flooding and climate change. In general, during Republican times, weather was cool and humid, yet pleasant. Just before and during the years of Imperial Rome, there was less rain, and temperatures were warmer (Lamb 1995). Between A.D. 800 and 1200, the climate was generally warm and dry, and salt water backed up into the coastal aquifers. This climatic trend began to change after 1200, and there was a very wet period between 1310 and 1320, accompanied by floods. During the "Little Ice Age" in Europe (approximately 1500 to 1800), Rome experienced lower temperatures, increased rainfall, and numerous floods of the Tiber (Bencivenga et al. 1995).

Garibaldi's Project to Divert the Tiber

Beginning in the time of the ancient Romans, proposals to control the Tiber's rage were numerous, including suggestions to reduce or eliminate flooding by creating better drainage, dikes along the riverbanks, and an actual diversion of the river around Rome. The first known proposal to divert the Tiber came from Caesar. A similar plan was put forth fifteen centuries later, in 1576, by Andrea Bacci, who proposed that floodwaters be directed around the east side of Rome, along the valley of the Aniene River, across the Prati di Castello, and through the Valle d'Inferno. Such diversion plans remained only curiosities, however, until the five-day flood of December 1870, when damage was in the millions of lire, many lives were lost, and misery was rampant. The capital of the fledgling Italian state was not prepared, and voices now were raised demanding that the government deal with the flooding problem. Standards for urban flood control had been set in such cities as London and Vienna during the 19th century, and citizens wanted to do the same in Rome. The engineer Alfredo Baccarini prepared a proposal to divert the Tiber by excavating a channel from the

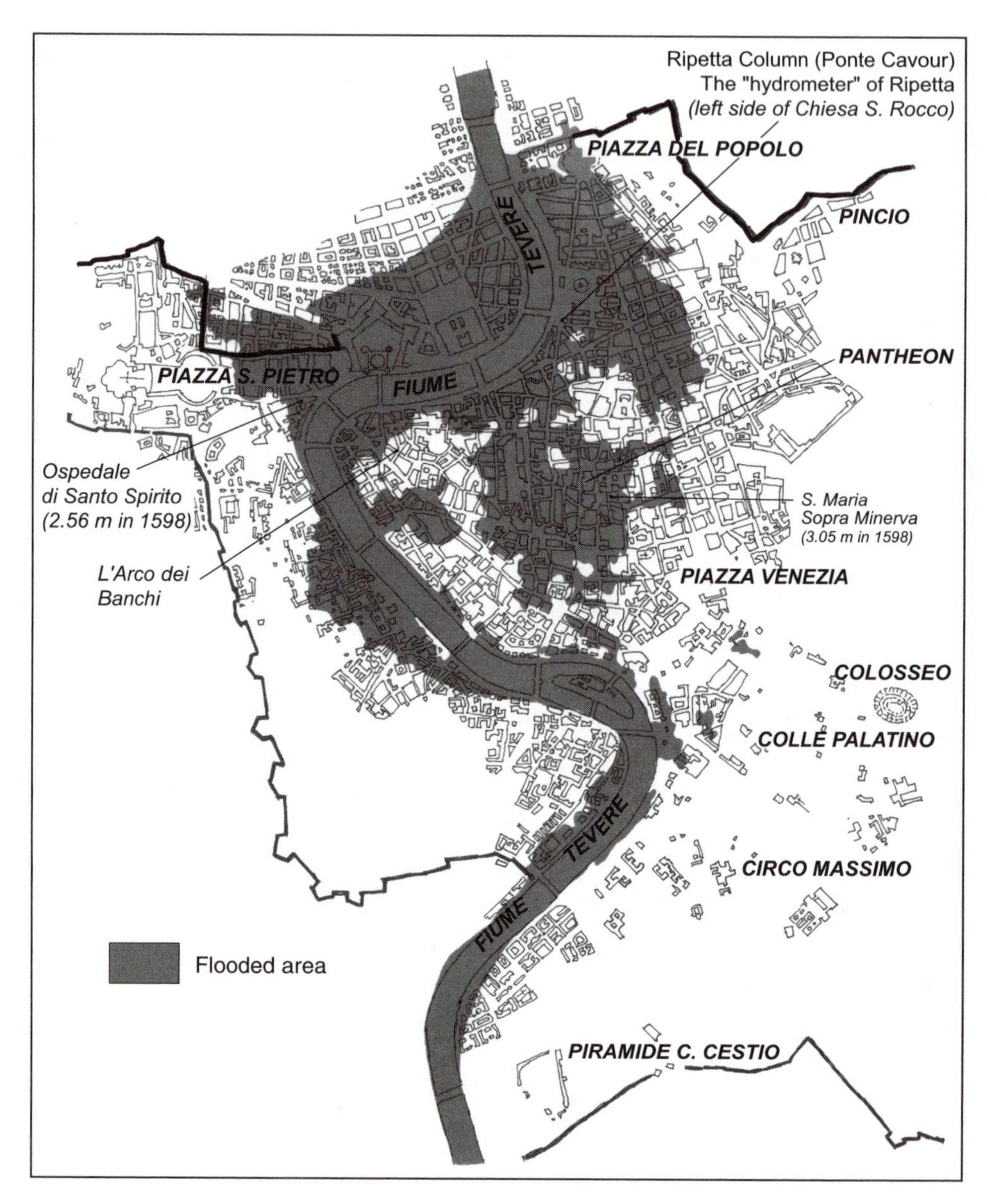

This map of central Rome during the flood of 1870 shows the extent of the floodwaters and marks the locations of many "high-water mark" plaques and columns. (Adapted from Bencivenga et al. 1995)

Aniene River valley, traversing the Varco di San Giovanni and crossing the Valley of Almone. The proposed channel would have diverted excess water during heavy rains and also provided a canal for irrigation and navigation.

As the previous year's flood faded into memory, evaluation of the proposal was delayed until 1873. At this time the concept was taken up by a new champion—General Giuseppe Garibaldi, father of a unified Italy. Garibaldi favored a new proposal by Paolo Molini, a Tuscan who enthusiastically supported a geologic study of the route before excavation, and Alessandro Castellani, a Roman, Garibaldino, industrialist, and archeological dilettante.

Garibaldi wrote to Castellani on November 12, 1872: "The winds of change in Rome include the world of science. Thousands of Italians deny the reputation that they are holding back the country. Finally we recognize the colossal advantages of your work" (from Morolli 1982).

Garibaldi's excitement was evident as he compared the project to those of the Suez and Panama canals. He believed that the diversion would liberate Rome from the floods. His dedication to the project was evident when he wrote in the paper *L'Opinione* on November 30, 1872: "I certainly don't take credit for the initiative to channelize the Tiber. I support the proposals of the scientists Castellani and Molini, who recommend continuation of the plan to bypass Rome, which will result in benefits for the citizens there" (from Morolli 1982).

In a letter to Silvio Spaventa, minister of public works, on December 29, 1874, Garibaldi stated: "It is my pleasure to inform you of the starting date for the study to divert the Tiber." At every opportunity, Garibaldi pushed the project, and with his endorsement it automatically was politically "blessed." The original plan was to excavate a grand canal from the "Bend of the Serpent," passing over the confluence of the Tiber and Aniene rivers, down the Valley of Marranella di Pietralata, cutting off the heights of San Giovanni near Porta Furba, and then descending via the Valle dell'Almone, toward San Paolo and the ancient bed of the Tiber. The plan's 17-kilometer-long (10.6-mile) canal would have had a gradient of 3 degrees. The diverted river would have emptied into a new harbor at Fiumicino surrounded by 2 kilometers (1.2 miles) of mole, to be designed by James Wilkinson, an Englishman.

In 1875, commissions created to evaluate the proposed diversion of the Tiber recommended a less ambitious plan that would be turned over to a private firm similar to the Suez Canal Company. In July of that year, the Chamber and Senate passed legislation that crippled the project. Not to be deterred by an intransigent government, proponents submitted a more modest "Second Project of Garibaldi," in which 1 cubic meter (264 gallons) per second of floodwater would be diverted into a 20-kilometer-long (12.4-mile) canal that would reenter the riverbed below San Paolo—the main purpose was to maintain a constant flow in the river. This less ambitious, second proposal was derided by political opponents as the "waste pipe."

Continued political infighting led to a meeting on November 27, 1875, between Garibaldi and the Superior Council of Public Works. Garibaldi's project was defeated, with nineteen voting against and three in favor of the "waste pipe" version. A political cartoon of the time likened Garibaldi to Gulliver and the politicians united against the grand project as Lilliputians. The political battles continued until March 18, 1876, when five possible methods were proposed to control flooding along the Tiber:

- constructing reservoirs above Rome,
- diverting the Tiber around Rome,
- installing a "waste pipe" for diverting excess rainwater,
- straightening the river, and
- constructing a system of higher walls along riverbanks throughout the city.

Although reservoirs were constructed and walls built along the river, none of the diversion plans supported strongly by Garibaldi were finalized. The last flood, in 1958, submerged part of the Isola Tiburina and no more. Dikes and walls erected along the Tiber, although they have degraded the natural appearance of the river, have been effective in reducing the flood hazard. Beginning in the early 20th century, dams (including twenty-three for hydroelectric generation) have been built within the Tiber basin; they have collectively reduced the flood hazard

"The new Italian Gulliver." This political cartoon (from *Il Pappagallo*, 1875, no. 7) shows Garibaldi held down by politicians of the Italian Chamber and Senate, who wanted to stop a proposed diversion of the Tiber around Rome he hoped would prevent the frequent floods that plagued the city.

along the Tiber but haven't eliminated the flash floods that can occur locally along its tributaries.

Although the engineers who proposed diverting the Tiber around the eastern side of Rome could not have known it, that rerouting would have mimicked nature. Recently, geologic mapping and data from numerous wells in the area of Rome have been used to reconstruct the course of the ancient Tiber before it was diverted by volcanic eruptions of the Alban Hills and Sabatini volcanoes. The course went south-southeast, from near the confluence of the modern river with the Aniene River to a point near today's Cinecittà movie studios, where the data sets end—beyond that point, the Tiber's early path is unknown. The variously proposed man-made routes would have followed nearly the same route as the ancient riverbed, which is now buried by sequences of volcanic rock.

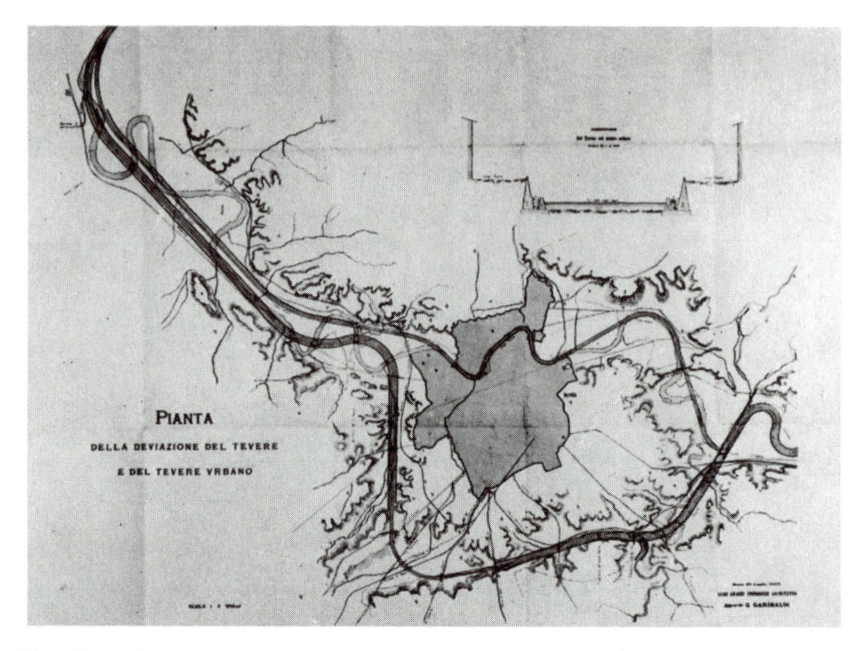

The diversion of the Tiber proposed by colleagues of Garibaldi (the bottom route) followed the natural course of the river as it existed before volcanic activity in the Alban Hills pushed it to the west. That course followed a path to the southeast, behind Rome, below what is now the suburb of Cinecittà (in this case, north is to the right). The plan failed, and the sinuous course above is today's route of the Tiber through Rome.

The Tiber Far Afield: Floods and the Decline of Roman Commerce

As a part of the modern-day commerce that fuels the Roman economy, international and domestic flights land at Fiumicino Airport every hour of every day. Jet-lagged and distracted passengers arriving on these flights probably aren't very interested in the fact that Fiumicino Airport is located partly on an ancient harbor, the Port of Claudius. This harbor, dredged in A.D. 54, was one of several harbors at the mouth of the Tiber that combined to make Rome a commercial power in the Western world.

One of the best ways to grasp the importance of ancient Rome's early seaports is to take the local train or bus toward the sea to Ostia Antica, another commercial port that is now 4 kilometers (2.5 miles) from

The ancient Roman port city of Ostia, which is now located 4 kilometers (2.5 miles) from the sea, was left landlocked by an ever-growing Tiber River delta. Ostia (now Ostia Antica), as a merchant port where materials were transshipped up the river by barge, was important to the economical and political health of Imperial Rome.

the sea. Ostia's well-preserved theaters, baths, multistory apartment houses, offices, and merchant warehouses give you a good feeling for ancient Roman life in what must have been a bustling, noisy mercantile port city. The population was perhaps 100,000, a very large city for the times. However, the city fell victim to history and the relentless growth of the Tiber delta. The decline began in the 4th century A.D. following a reduction in trade, gradual silting of the harbors, and the spread of malaria, and busy Ostia was gradually abandoned to its now quiet site inland on the Tiber's delta.

What Is a River Delta?

All rivers carry a sediment load: pebbles that scuttle along the riverbed and fine sand and silt that are suspended in the flowing current. The finer-grained sediment is carried as long as the water continues to move, but when a river reaches a shoreline, as when the Tiber reaches the Tyrrhenian Sea, much of the suspended sediment falls to the bottom in the still water and is thus "deposited." Some of the sediment is carried along the shore by currents and wave action, which feed and maintain beaches. Continual sedimentation at a river mouth eventually builds a *delta* (a term derived from the flattened triangular shape, with the river entering at the apex) that continues to expand into the sea. The growth and shape of a delta depend, in part, on the size of the river and its sediment load. In the case of the Tiber, however, a rising and falling sea level also affected the delta's growth. Over the last 13,000 years, the Tyrrhenian Sea has risen 67 meters (220 feet) to its present level. When the sea level was much lower, the Tiber was cutting a deep channel through what is now Rome, so the delta was also much lower and closer to Rome.

You probably won't be able to see the Tiber delta unless you are looking at a map or satellite image. The distance from the coast to where the river debouches from the hills that flank it is 13 kilometers (8 miles). The delta has 30 kilometers (18.6 miles) of coastline and an area of 150 square kilometers (58 square miles); below sea level, the delta covers an additional 500 square kilometers (193 square miles). All the flat land along the Tiber between southwestern Rome and the Fiumicino Airport is on the delta, where it is a mélange of swamp, old sand dunes, reclaimed farmland, and slow-moving drains or creeks.

To unravel the Tiber's complex history and understand the growth of the delta, you must think of it in terms of four dimensions: length, width, depth, and time. A team of geologists, geophysicists, and historians from the University of Rome–La Sapienza, led by Piero Bellotti, recently studied the Tiber delta's history, using data from 226 wells, sea-floor samples, geophysical surveys, and archeological information from the last 2,500 years. They concluded that the delta is a mixture of deposits left by the river and coastal plain over the last 800,000 years, and many of the lateral and vertical variations in the deposits are linked to the rise and fall of sea level. The Tiber delta that we see today is part of the most recent depositional sequence (over the last 100,000 years). Carbon-14 dating has added detail about the last 13,000 years, a time of gradual sea-level rise.

We are fortunate also to have written records from the 3rd century B.C. to the 4th century A.D. to chronicle the delta's steady forward march. Roman writers, including Pliny the Elder and Plutarch, described the Tiber of their time as a calm river with few floods; they speak of a harbor city at its mouth (Ostia), which was founded in the 4th century B.C.

During the 1st century B.C., when the Tiber rarely flooded, Julius Caesar intended to reclaim the coastal area not connected with the river, which at the time had begun to silt up. Claudius eventually constructed a harbor north of the river mouth between A.D. 42 and 54, and Trajan enlarged it by adding a hexagonal basin in A.D. 110, creating what was the largest man-made harbor in the world. The harbor was connected to the Tiber by a canal (the "Fiumicino") that eventually gave the river a second mouth; this change has since modified the delta's shape. Pliny the Younger, who was active in the late 1st–early 2nd century, owned a villa near Ostia that overlooked the sea; today, the ruins of his villa are several hundred meters from the sea and provide mute evidence of the delta's growth.

Information about the delta is sparse for the time after the fall of the Roman Empire (A.D. 476). Papal bulls of about A.D. 1000 note that Claudius's harbor was partly filled with silt, the harbor of Trajan was a lake, and the Tiber now had a second, well-defined mouth. Beginning in 1420, the popes constructed watchtowers at the river's mouth; by 1773, a sequence of six had been built to protect the continually expanding shoreline.

Historians and scientists have studied the Tiber delta since the 18th century. Its growth has been the main focus, but since the 1960s, research has also evaluated the effect of dams on the Tiber, which have reduced the river's sediment load, slowed delta growth, and enhanced beach erosion. Reforestation, reclamation of coastal marshes, and quarrying the riverbed for sand and gravel have also affected sediment volume.

Bellotti and his coworkers established a link between floods on the Tiber and the delta's geologic history by showing that a period of rapid delta growth during the 15th through 19th centuries corresponds to a period when frequent floods afflicted Rome, and Europe experienced the "Little Ice Age."

• • •

As you wander through the maze that is Rome on the floodplain, watch for the many escapes to higher ground via streets that follow the now well-hidden tributaries of the Tiber. These tributaries were once intermittent streams, sometimes requiring a boatman to cross, and some were stagnant, poorly drained marshes.

CHAPTER **6**

The Tiber's Tributaries in Rome

CLOGGED WITH HUMANKIND'S DEBRIS

THE MAJOR STREETS Via del Tritone, Via Barberini, Via Vittorio Veneto, Via Cavour, Via di San Gregorio, Via delle Terme Caracalla, and Via Labicana all rise into the seven hills of Rome along now-buried tributaries of the Tiber River. One of the tributaries, the Aquae Sallustianae, which was fed by the Sallustiane springs, flowed between the Pincian and Quirinal hills (a small drainage now followed by the Vie del Tritone, Barberini, and Vittorio Veneto) and into a swampy area used for grazing goats (called, not surprisingly, the "Goat Marsh"). Between the Viminal and the Esquiline hills was another stream whose waters passed between the Capitoline and Palatine hills, through the swamps of Lacus Curtius and Velabrum Minus, and finally into the Tiber. These tributaries that cut into the tuffs of the plateau now are partly filled with alluvium and man-made debris, which are important factors in the city's past, present, and future. As you traverse Rome, you're treading on layers upon layers of debris, most of which is tastefully overlain by buildings and pavement.

Going west down the Via Labicana toward the Colosseum, you are trekking along what was a swamp between the Esquiline and Celian hills. If you follow the Aurelian Wall west-southwest from San Giovanni in Laterano to the Baths of Caracalla, you are passing along the route of yet another stream that fed the swamp in the Valle delle Camene, which was eventually drained to become the Circus Maximus. At the end of the 3rd century B.C., Romans had to cross the substantial swamp by ferryboat. The Romans dug channels for the streams to eliminate the large, unhealthy swamps and, eventually, to create the water collection network called the Cloaca Maxima (the "Big Drain").

The Cloaca Maxima, still visible along the eastern bank of the Tiber near the Ponte Palatino, was ancient Rome's main storm sewer. It was

The Pyramid of Caius Cestius, a wealthy Roman magistrate buried here in 12 B.C., was incorporated into the Aurelian Wall nearly 300 years later. The land southwest of the pyramid became the Protestant Cemetery, which includes the graves of such dignitaries as the poet John Keats, Julius Augustus, the illegitimate son of Goethe, and Antonio Gramsci, the first leader of the Italian Communist Party. As you pass through the busy intersection of the Piazzale Ostiense, the pyramid and adjacent ruins originally built to drain the central part of the city, especially during heavy rains that produced flash floods. Reputedly begun by the Etruscan king Tarquin the Proud to drain the tuff plateau through the Velabrum and Argiletum valleys, the drain was an open canal until the 3rd or 4th century B.C., when Roman engineers began to cover it with stone arches. Like many such structures in the city, the Cloaca Maxima was built of durable tuff (*Tufo pisolitico, Tufo Giallo della Via Tiberina*, and the *peperino* of Gabii, all discussed in chapter 3). These semicircular arches are nearly 5 meters in diameter and are still intact after more than 2,000 years.

The Tiber's tributaries, which once so clearly delineated the seven hills of Rome, obviously are not so easily seen in the 21st century—in part because of the closely spaced buildings throughout the city, but chiefly because the streambeds have been so well masked by man-made debris.

Debris: Engineer's Curse and Archeologist's Treasure

Humans generate and accumulate a lot of waste. We sack it, stack it, hide it, burn it, recycle it, or pay someone to take it elsewhere—anywhere we can't see it or smell it. In today's world, with greater material consumption, we generate increasingly enormous masses of debris that must be disposed of or recycled. Over the last 3,000 years, surely the debris accumulating in Rome was equally undesirable, but it is now eagerly excavated by archeologists in their search for clues to the past. Wherever the written word hasn't survived, we depend on accumulated waste for information about a community's lifestyles and infrastructure. You can visit the Roman fora and watch archeologists avidly digging through both the layers of debris left by everyday life and the major "marker beds" that are the recognized products of invasion, fire,

---

appear to be in a hole; actually, they have not sunk but were slowly surrounded by rising debris left by Rome's residents and catastrophes like the fire of A.D. 64. Layers of debris in the area of Piazzale Ostiense are 5 to 10 meters thick, and the base of the pyramid is 3 to 4 meters below today's streets.

demolition, and reconstruction. If you look carefully, you'll observe the complexity and heterogeneity of the layers and appreciate the years of education and experience that lie behind archeological observations.

We know this region has been occupied for at least 14,000 years, although it is difficult to identify Paleolithic and Neolithic sites below the younger debris of the city. Stone tools, pottery, and (eventually) copper weapons found in and around Rome are evidence of the first 10,000 years of human occupation. The copper most probably came from mines in Tuscany, near Monte Amiata. Obsidian tools likely had sources in Sardinia or the volcanic islands of Lipari and Palmerola. During the Bronze Age (approximately 2300 to 1000 B.C.), residents occupied multiple sites along the Tiber (as revealed in excavations at Sant'Omobono, Cisterna, and Veii). Throughout the time of Republican and Imperial Rome, with its population then reaching a million, continuous construction of new buildings, greater demand for imported consumer goods (including building stone and oil containers), and the generation and burial of waste substantially increased the rate at which the city rose above its original geologic foundation. This process has changed in the last few decades because debris is being hauled into the adjacent countryside, a practice that will undoubtedly confuse future generations of archeologists.

Geologists are interested in the debris left by previous Roman residents because it has modified the terrain. There have been so many human-caused changes that one of Rome's seven hills—the Viminal—is all but invisible to the casual observer. Throughout the historical center of Rome, nearly everything is covered by at least 2 to 5 meters (6.6 to 16 feet) of debris. On the alluvial plain in the neighborhoods of Trastevere and Pigna, the debris is 5 to 10 meters (16 to 33 feet) thick; along what were tributary streams flowing into the Tiber, the debris layers are 10 to 15 meters (33 to 49 feet) thick. Within the limits of the Aurelian Walls, there are 93 million cubic meters (121 million cubic yards) of man-made debris. This is a lot of material! However, the citizens of Rome have had 3,000 years during which they slowly collected the waste, buried or mounded it, and modified the natural terrain. The accumulation has not been a steady process; it has fluctuated from ordinary daily trash accumulation to massive layers produced by earth-

In archeological excavations in the Roman Forum, it is clear that the city has evolved atop its own debris; in this case, we see rubble from Republican Rome through to modern constructions such as the Curia, visible behind the Temple of Saturn on the right. The Curia was the chamber of the Roman Senate.

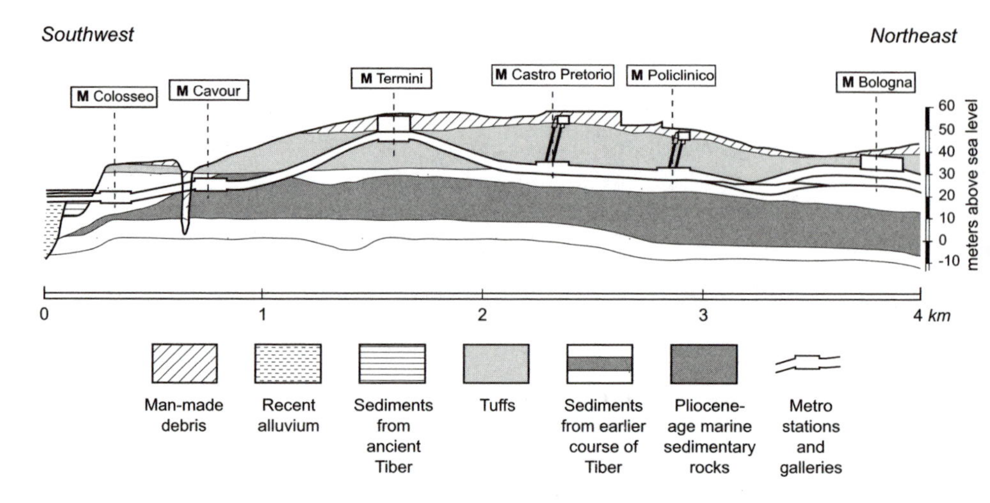

You get an idea of the remarkably thick layer of man-made debris that covers the city by studying this geologic cross section along the Metropolitana B (subway) between stations at Piazza Bologna and the Colosseum. For example, the ravine near Cavour station is filled with 20 meters (66 feet) of debris. The illustration is exaggerated vertically to show the relations between geologic units (measured in meters above sea level).

quakes, fires, invasions, or large-scale renovations of the city by leaders like Mussolini.

On a first visit to Rome you might be forgiven for thinking that nearly all the major structures were built in holes. The Church of San Vitale, next to the Palazzo di Esposizioni on the Via Nazionale, is 3 to 4 meters (10 to 13 feet) below street level. The base of the Pyramid of Caius Cestius, or the Piramide, along the Via Ostiense at Porta San Paolo, is also 3 to 4 meters below the numerous buses and cars that roll past this unique monument. Even the base of Trajan's Column and most of the buildings in the Roman fora rise out of depressions. Be reassured—these monuments and churches, ranging in age from the 1st century B.C. to the 16th century A.D., are not sinking; they have been slowly engulfed by the accumulated debris of an evolving city.

Deposits more than 15 meters (49 feet) thick are located where builders piled the debris from construction, industry dumped its refuse, or locals simply wanted to fill in an annoying ravine. Familiar features of Rome that overlie thick debris layers include the Piazza Barberini, Ter-

The small Church of San Vitale, near the Palazzo di Esposizioni on the Via Nazionale, sits well below street level, surrounded by the debris that has accumulated since medieval times.

mini Station, the southwest embankment of the Circus Maximus, Via Cavour, Porta Melonia, and Piazza Tuscolo.

The winner for thickest debris deposit in Rome is the Monte Testaccio, a 250-by-180-meter (820-by-590-feet), 36-meter-high (111-foot) hill in the Testaccio neighborhood. The volume of this man-made hill is 1.6 million cubic meters (2.2 million cubic yards)! The hill was a dumping ground for warehouses and workshops in the "emporium zone" along the Tiber River south of the Aventine Hill. The Emporium consisted of a river port and warehouses constructed with tuff blocks that covered an area 467 meters (1,530 feet) long and 60 meters (197 feet) wide. Much of what arrived at the river port first went to warehouses and workshops in what is now the Testaccio neighborhood.

Monte Testaccio is composed of mostly broken amphorae (terracotta jugs) used to import oil from around the Mediterranean. Rodri-

This aerial photograph shows the area of the Monte Testaccio, a 36-meter-high (111-foot) hill of debris, mostly broken amphorae accumulated between about A.D. 145 and 255. Imperial Rome's major river port was nearby, and the neighborhood around Monte Testaccio was an industrial district.

guez-Almeida, in his analysis of Roman trade based on the components of the hill, estimated that the number of broken amphorae totaled 53,359,800, which would have contained 37 million cubic meters (nearly 10 billion gallons) of imported olive oil! Even averaging this volume of oil over the 110-year lifetime of the landfill, the per capita

The Monte Testaccio is a man-made hill now surrounded by residences and restaurants.

annual consumption would have been 34 liters (9 gallons), assuming a population of 1 million. This estimate seems rather high, but we must remember that oil was used for lighting and cosmetic purposes, as well as for cooking. Other debris components include chunks of pozzolan concrete, plaster, broken roof tiles, bits of stone pavement, and fragments of glass and ceramic oil lamps—much like modern landfills. After the fall of Imperial Rome, the Testaccio Hill went virtually unnoticed until the 18th and 19th centuries, when it became a tourist attraction. Today, the ancient industrial zone is becoming a popular site for restaurants and nightclubs.

The Palazzo Valentini, which houses the offices of the Provincia di Roma, is located immediately north of Trajan's Column, east of the Piazza Venezia, and is a potential victim of history and gravity. Follow along as we begin a virtual ascent through the complex and numerous layers beneath the Palazzo Valentini. The debris sequence begins with the Temple of the Deified Trajan, built by Hadrian around A.D. 117 on a substantial stone foundation. Nearby were Trajan's Forum (the largest of the Imperial fora), and Trajan's Column. The temple fell into disuse after the fall of the Roman Empire, and the site became a rubble pile. From the 5th to 12th centuries A.D., the temple was smothered by one of Rome's medieval quarters. Between 1583 and 1585, this neighborhood was demolished to build the Palazzo Borrelli, which was never completed. Later construction (A.D. 1650 to 1689) extended the palazzo toward Trajan's Column, and in about 1740, the building became the Church of Santissimo Nome di Maria. It came under new ownership in 1750 when the Imperial cardinals carried out work to restore the library. Between 1796 and 1830, the new owner, Vincenzo Valentini (thus, its current name), added a new section that had a view of Trajan's Column. The Provincia di Roma took over the palazzo in A.D. 1874, remodeling it and adding another level. In the mid–20th century, part of the building was turned into a bomb shelter.

To sum up, the man-made debris under the palazzo, representing the ebb and flow of Roman history, is between 6 and 16 meters (20 and 52 feet) thick. The heterogeneity of the debris deposits results in irregular compaction and subsidence under the weight of the palazzo's foundation. The difference between the compaction of unconsolidated medieval and Renaissance debris layers and the rock foundation of the Temple of the Deified Trajan has led to cracking and even tilting of sections of the overlying palazzo. The palazzo is not the only important building that is threatened in this manner.

In August 1969, in the Palace of Justice (Palazzo di Giustizia di Roma, or Il Palazzaccio), a granite corbel collapsed, fell through the ceiling, and landed in a ground-floor hall. Located near Castel Sant'Angelo between the Tiber and the Piazza Cavour, the palace was built in 1893 directly over a spring on the right bank of the river. The corbel fell because differential subsidence of the palazzo's foundation was tearing the building apart. When it was examined in 1970, the palazzo had more than 450 cracks on each of its three levels. Precise leveling surveys made after the cracks were mapped revealed that the building was subsiding at about 5 millimeters (one-fifth of an inch) per year. That alone wouldn't be catastrophic, but the subsidence is not uniform below this enormous building. The facade nearest the Piazza Cavour had sunk 50 centimeters (20 inches) between 1893 and 1970, but in more recent times, most of the subsidence has occurred on the opposing façade, nearer the Tiber.

We know that cracking and tilting of parts of the palace are caused by differential compaction again, but in this case, by unconsolidated river deposits (sand, gravel, and clay) and man-made debris (including the remnants of Imperial Roman ruins) underlying the palazzo; the deposits are not compressed equally under the weight of the stone palazzo, and some deposits sink more than others. Building over a spring without considering the effects of water-saturated sediment also added to the problem. Furthermore, engineers investigating the stability of the palazzo found that the foundation itself lacked rigidity; thus, it could not "float" as a single block on the underlying sedimentary and debris deposits.

In general, early Roman engineers constructed the city's buildings on stable foundations. There were exceptions, however, such as the Colosseum, which was constructed over a contact between alluvium and the rock of the Oppian Hill. Even just after medieval times, builders were already facing the difficulties of laying foundations on heterogeneous debris layers accumulated throughout earlier Roman history. Today, most of the younger buildings suffer only from annoying cracks, but the effect of differential subsidence can lead to the more serious problem of tilting, which requires repair and retrofitting.

Difficulties associated with the lateral and vertical heterogeneity of both man-made debris deposits and unconsolidated alluvium have provided a key to understanding another problem: the way Rome's buildings and monuments respond to earthquakes, which are so common in Italy.

Geologists Evaluate the Risk of Earthquakes

To evaluate the earthquake risk to man-made structures, we need to look beneath, at the rock or sediment type as well as the thickness and shape of each deposit. We need a geologic evaluation and, if we are lucky enough to have access to sophisticated computer databases, some numerical modeling of the site in order to estimate any strong ground motion that would accompany a future earthquake. Structures built on rock have the greatest chance for survival. Those built on deposits of poorly consolidated alluvium, especially in narrow valleys, are subject to amplified ground motion and thus have an increased risk of serious damage. We can also extract earthquake histories by digging trenches across faults to look at indicators for ground movement and by collecting material for radiometrically dating sediment layers left by past earthquakes.

We must not only determine the rock types (usually from local outcrops) but also understand something about their geometry and distribution. Given that old cities like Rome have covered most of the outcrops with man-made debris, it is difficult to evaluate what underlies the city unless you have a large drilling budget that allows you to poke holes through the debris. Geologists can use information about damage to historical structures caused by past earthquakes to predict zones where buildings are vulnerable to strong ground acceleration. Compilations of the time and degree of damage from earthquakes will also help us develop the data sets we need to estimate future risks.

Earthquakes in Rome?

A tablet located in the entrance of the building that has become a symbol of Rome and the Roman Empire—the Colosseum—describes the generosity of Decius Marius Venantius Basilius in subsidizing repairs after earthquakes that occurred between A.D. 443 and 484. Rome is subject to both small and large earthquakes, and the chronicle of these events is one of the best in the world because of Rome's long historical record. The monuments of Rome are proud survivors of seismic activity, but geophysicists and conservators alike are concerned about the potential damage from future earthquakes.

The Flavian Amphitheater—better known as the Colosseum—is easily recognized by people everywhere because of its asymmetry: the northern half has four levels of arches and looks complete, whereas the southern half has two and one-half to three levels remaining and is obviously damaged. This damage appears to have been caused mostly by earthquakes over the last 1,900 years and rather less by citizens and builders "mining" the structure for construction material.

One of the authors of this book (Funiciello) and the geophysicist Antonio Rovelli of the Italian Institute of Geophysics have studied earthquake effects on major Roman monuments, including the Colosseum. They discovered that damage to the amphitheater by the many earthquakes that have hit Rome during the last two millennia was related in large part to the underlying geologic deposits. Roman engineers were superb, but they did not consider the basic issue of geologic underpinnings before building this arena. This failing is not unique; construction today in most of the world's cities suffers from the same problem: a lack of understanding of the geologic foundation below (or, in cities with hills, the geology above).

Using data from earlier geologic mapping and a series of exploration drill holes, Funiciello and Rovelli found that the Colosseum had been constructed across the boundary (in geologic terms, the *contact*) between Pleistocene age sedimentary rocks and tuffs (volcanic deposits from the Alban Hills) and the unconsolidated alluvium of a creek that ran between the Palatine, Esquiline, and Celian hills and then into the Tiber through what is now the valley containing the Circus Maximus. Adding further instability, a small valley once dammed to form an artificial lake was, in turn, covered with a layer of burned debris left by the great fire of A.D. 64 (carbonized wood from the fire was discovered during geo-engineering drilling). Vespasian, the first emperor of the Flavian family, who succeeded Nero, decided that the small valley was the perfect site for a great arena, which he had constructed in little more than five years (his son Titus dedicated the arena in A.D. 80).

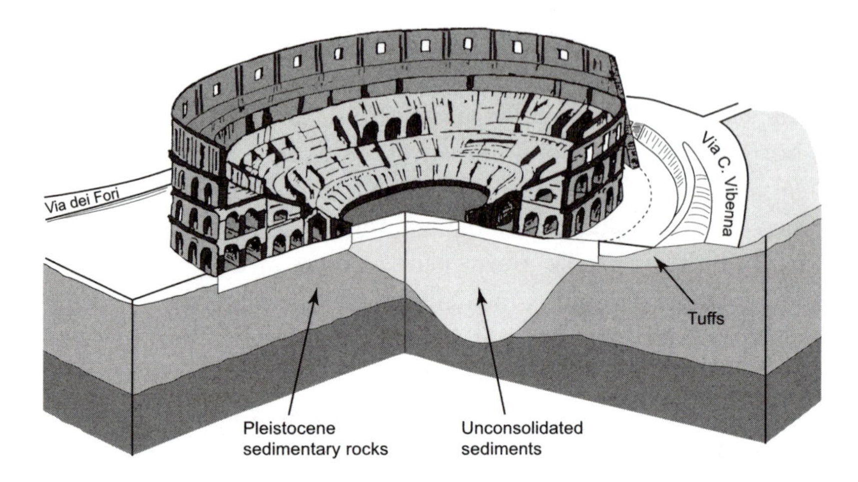

As you can see from this diagram of the Colosseum (Flavian Amphitheater), the north side of the Colosseum (to the left), which is underlain by marine sedimentary rocks and tuff, suffered only light damage during the earthquakes of the centuries. The southern side, overlying an ancient creek filled with poorly consolidated sediment, was severely damaged by excessive ground acceleration within the alluvium during large earthquakes.

Despite its massive, 13-meter-thick (43-foot) concrete foundation, the variation in rock types under the stadium becomes evident when one looks at structural damage caused by a few large earthquakes. After A.D. 484, the Colosseum was gradually abandoned as a site for sporting activities; it was used by criminals as a refuge and by others as a corral for cattle. During the 13th century, it became a fortress for a wealthy family. The debris left after earthquakes served for a while as an ersatz "quarry" before Pope Benedict IV saved this landmark by consecrating it.

Using the results of geologic drilling to visualize the three-dimensional framework of the site, Funiciello and Rovelli simulated the ground acceleration that would have occurred below the stadium during an earthquake. In the unconsolidated stream sediments of the tributary underlying the southern half of the Colosseum, the ground acceleration was strongly amplified, enhanced by the shape of the valley. Large earthquakes heavily damaged the portion of the Colosseum underlain by a sediment-filled creek bottom, but adjacent portions located on rock suffered only light damage.

Basic data for an earthquake history should include the degree of damage to buildings, bridges, and monuments; however, not every historical record indicates if an earthquake was responsible for property damage. In ancient Rome, buildings often collapsed, even without earthquakes. During Imperial times, Emperor Augustus limited the height of private buildings to 21 meters (70 feet), and anyone who wished to build above that height needed the emperor's permission. In the following century, Trajan further limited construction to about 18 meters (60 feet). These regulations were prompted when a period of enormous demand for low-cost housing resulted in less-than-excellent projects. Tall tenements frequently collapsed because of poor construction, but their destruction was assured during earthquakes. As is often the case in the modern world, cheap housing provided great profits but put the residents at risk. The writer Juvenal, who died around A.D. 130, lamented the unstable, badly constructed buildings of Rome: "We live in a city that is supported, more or less, by props." Some Italians agree, saying that not much has changed, especially with the apartment houses put up quickly after World War II in response to a postwar housing crisis.

In early documents, earthquakes were lumped into the same category of disaster as war, revolt, and invasion. Dates for earthquakes that occurred before A.D. 1000 are not precise, especially for events between A.D. 500 and 1000. The first recorded description of a Roman earthquake was by Livy, writing in the final decades of the 1st century B.C., who mentions the tremors of 461 B.C.: "The ground was shaken by a violent earthquake." Livy, in this terse account, described the event in the context of miracles that occurred that year. Not much was said about actual damage to the city except for the falling of large objects. An earthquake in 83 B.C. damaged public buildings and houses, and Appian (2nd century A.D.) interpreted the event as a portent for civil war (this is certainly plausible—the same happened in Managua, Nicaragua, in 1972). Other significant earthquakes noted by Roman writers occurred in 179 B.C., about 71 B.C., and A.D. 15, 51, 443, 484 or 508, 801, and 1091.

One of the most damaging earthquakes affecting Rome occurred on September 9, 1349. Near the epicenter in the central Apennines, the intensity was Mercalli grade X; in Rome it was grade VIII. The poet

TABLE 6.1
Large Earthquakes That Caused Damage in Rome from the 13th Century A.D. to the Present

Zone of Epicenter	*Year*	*Intensity, Epicenter*	*Intensity in Rome*	*Comment*
Umbria-Marche Apennines	1279	IX	V	—
Central Italy	1349	X	VII–VIII	Serious damage to the basilicas and medieval towers
Aquilano	1461	X	V	No description
Umbria Apennines	1703	XI	VII	Serious damage to many buildings
Aquilano	1703	X–XI	VII	Serious damage
Abruzzo Apennines	1706	X–XI	V	
Alban Hills	1806	VIII	V	Minor damage to some buildings
Roman area	1811	VI	V–VI	
Roman area	1812	VI–VII	VI	Partial collapse of many buildings
Alban Hills	1892	VII–VIII	V	Minor damage
Roman area	1895	VII	V–VI	Some serious damage
Alban Hills	1899	VII	VI	Some serious damage
Roman area	1909	VI	IV–V	Minor damage in Monte Mario
Alban Hills	1911	VI	IV–V	Some minor damage
Marsica	1915	XI	VII	Much damage and some partial collapse
Roman area	1919	V–VI	V	Some minor damage
Alban Hills	1927	VIII	V	Some minor damage
Val Nerina	1979	VIII	V	Some minor damage
Roman area	1995	VI	IV–V	Some minor damage
Umbria-Marche Apennines	1997	VIII–IX	V	Some minor damage

Source: Donati, Funiciello, and Rovelli 1998.

Petrarch, who was in the city for the Jubilee of 1350, found the city "prostrate," citing severe damage to the structures most frequently visited and admired by pilgrims, including bell towers and basilicas. Worried about the expected influx of pilgrims, Pope Clement VI concentrated on repairing damage to the most important basilicas: San Paolo,

TABLE 6.2
The Mercalli Earthquake Intensity Scale

Scale	*Intensity*	*Description of Effect*	*Corresponding Richter Scale*
I	Instrumental	Detected only by seismographs.	—
II	Feeble	Some people feel it.	—
III	Slight	Felt by people at rest, like a large truck passing.	Less than 4.2
IV	Moderate	Felt by people walking. Loose objects on shelves rattle.	—
V	Slightly strong	Awakens sleepers. Church bells ring.	Less than 4.8
VI	Strong	Trees sway. Suspended objects swing. Objects on shelves fall off.	Less than 5.4
VII	Very strong	Mild alarm among people. Walls crack. Plaster falls.	Less than 6.1
VIII	Destructive	Moving cars cannot be controlled. Chimneys fall, and masonry is fractured. Poorly constructed buildings are damaged.	—
IX	Ruinous	Some houses collapse. The ground cracks, and pipes break.	Less than 6.9
X	Disastrous	There are many ground cracks. Many buildings are destroyed. Some liquefaction of ground and many landslides.	Less than 7.3
XI	Very disastrous	Most buildings and bridges collapse. Roads, railways, pipes, and cables are destroyed.	Less than 8.1
XII	Catastrophic	Total destruction of man-made structures. Trees are torn out of the ground. The ground rises and falls in waves.	Greater than 8.1

Saint Peter, and San Giovanni in Laterano. This may have been the same earthquake that caused major damage to the Colosseum, leaving it half in ruins, as we see it today.

Seismic historians have had better documentation since medieval times for evaluating the earthquake intensity and subsequent damage of historic tremors.

The Earthquake of January–February 1703

> A series of earthquakes, causing violent shaking and notable damage, terrorized the residents of the city. This violent earthquake originated in the Apennines of Umbria and Abruzzo and was possibly the most important earthquake in the history of central Italy. Ground motion, occasionally with intensities of MCS grades IX and X, destroyed numerous towns, left thousands of victims, and produced ample evidence of effects on the landscape and groundwater.
>
> The people of Rome living in areas where the most severe ground motion occurred [about MCS grade VII—on the Tiber's floodplain and tributaries], were found in a grave state of terror and exhaustion, continually rebuilding and adding supports to structures and putting up notices of death and destruction. All spent nights in the open during the bad winter weather and not in the buildings.
>
> On the first day of March, 1703, an indication of the climate that reigned in the city was the band of criminals who posted a notice predicting the imminent fall of the city, to encourage the inhabitants to abandon their homes and then to rob them. The general fear led to the numerous religious functions celebrated in Rome during the following year. (Molin and Guidoboni 1989)

The ground motion in Rome during this event caused serious damage, mostly on January 14, and was also disastrous near Norcia; on February 2, the quake caused catastrophic damage in the city of L'Aquila and further damaged Roman buildings weakened during the first earthquake. Damage recorded in Rome included the collapse of a house near Santa Prassede and destruction of a loggia parapet near the "Quattro Fontane," which killed two brothers. Despite their solid appearance, town walls partially collapsed. Roof gables fell in Trastevere, and deep fissures opened in the walls of many public buildings. Many Roman monuments were damaged; particularly hard hit was the Colosseum, where three arches of the second enclosure on the south side were ruined (facing the Church of San Gregorio). The effects on groundwater, as described by Molin and Guidoboni, were notable in many of the city's wells ("For a while, the water was turbid and had a bad odor. Pressures in water systems dropped") the result of disrupted springs and broken pipes.

The most serious Roman earthquake of the 20th century occurred on January 13, 1915. The epicenter was located 80 to 100 kilometers (50 to 60 miles) north of Rome, in the Lazio-Abruzzo Apennines, where the intensity was Mercalli grade XI. This quake was felt throughout Italy and in parts of Yugoslavia. All wards and quarters of Rome were affected, although the extent varied. Most of the damage occurred in structures located on the alluvium of the floodplain and tributaries, and it was severe in all older floodplain neighborhoods, including Testaccio and Prati. The earthquake crumbled parts of the Aurelian Wall near the Porta del Popolo and Porta Maggiore, and near Porta Furba a part of the Claudian aqueduct was destroyed. Serious damage occurred in the churches of Sant'Agata de Goti and Santa Maria della Scala, the campanile of Sant'Andrea delle Fratte, and the cupola of San Carlo ai Catinari.

Sources of Earthquakes That Affect Rome

The earthquakes felt in Rome historically originate in three areas:

- Within a 15-kilometer (9-mile) radius of the city center—These earthquakes have magnitudes of less than Mercalli IV and shallow epicentral depths; they are rarely felt and are detected only by sensitive seismometers.
- Within "the Roman area"—Frequent earthquakes in the Alban Hills volcanic field have magnitudes of about Mercalli V; their cause is often attributed to either an influx or cooling (and shrinking) of molten rock located at great depths below the volcanic field. Along the Tyrrhenian coastline, infrequent earthquakes, with magnitudes of about Mercalli V to VI, are rarely felt within the city.
- Within 60 to 130 kilometers (37 to 81 miles) of Rome—Seismically active areas of the central Apennines have been the sources for all major earthquakes in central Italy. The larger quakes produce maximum intensities of Mercalli VII to VIII in Rome. Earth-

TABLE 6.3
Earthquakes and the Great Basilicas of Rome

Basilica	*San Giovanni in Laterano*	*San Paolo fuori le Mura*	*Saint Peter's (Vatican)*	*Santa Maria Maggiore*
Geologic Foundation	Tuff plateau	Across a boundary between tuff and alluvium	Complex overlap of Plio-Pleistocene marine sedimentary rocks, tuffs, and alluvium	Tuff plateau
Degree of Damage and Year	Roof collapse (9 September 1349) Varied damage to annexes (March 22, 1812) Plaster fell (July 7, 1899) The statue of Saint Paul fell from the facade (January 13, 1915)	Serious damage that required radical reconstruction of the roof, floors, and outside doors (April 29, 801) Campanile collapsed (sited on alluvium) and serious roof damage (September 9, 1349) Damage to clock walls (March 22, 1812) A capital fell from a column (November 11, 1895) Damage to the facade and apse. The marble cross over the entrance fell. Damage to the mosaics (January 15, 1915)	Unspecified damage (September 9, 1349) Light damage to the cupola and falling plaster (February 2, 1703) Some damage to the vault (March 3, 1812) Accentuated previous damage to the cupola lantern (January 11, 1895) Light damage and re-opening of older cracks and falling plaster (January 13, 1915)	Light damage to the vault (March 22, 1812) Unspecified damage to the apse (August 31, 1909)

Source: Data from Molin et al. 1995.

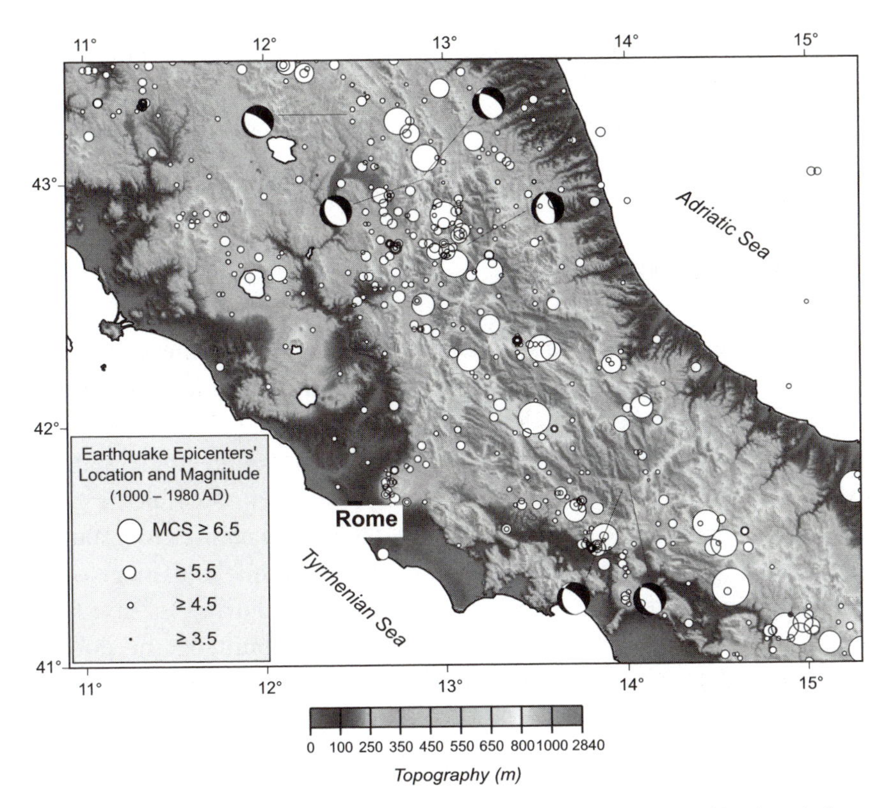

Looking at the epicenters of large earthquakes in the Italian Peninsula between A.D. 1000 and 1980, we can see that those of most of the large earthquakes that have damaged Rome are in the nearby Apennine Mountains, whereas few really damaging earthquakes originated along the coastal plain or in the volcanic fields that flank Rome. "MCS" refers to the Mercalli Cancani Sieberg earthquake intensity scale. (Historical and instrumental data are from Camassi and Stucchi 1997)

> quakes with intensities of greater than VII occur about every 500 years; those with intensities of VI to VII occur, on average, every 200 years.

Geologists have both observed and inferred faults below Rome that pose little risk to the city. The greatest hazard stems from the large earthquakes that originate in the nearby Apennines, especially when ground motion is amplified within alluvial deposits such as those found below the Tiber's floodplain. Rome's archeological monuments offer some insight into the degree of damage to a building during a single earthquake, which is related to the monument locations—on rock or alluvium.

The Aurelian Wall

The Aurelian Wall, Rome's defensive perimeter over the centuries, crosses nearly all the geologic units and terrain of Rome. One of the authors of this book (Funiciello) and his colleague I. Leschiutta used the Aurelian Wall to interpret the importance of geologic setting when calculating risk to man-made structures. They assembled a geologic cross section along the wall and used historical records to evaluate damage to both the wall and its gates. Most earthquake damage over the centuries has been to the Porta Metronia and Porta Ostiense (today's Porta San Paolo), where foundations were built on the poorly consolidated alluvium deposited in deep channels cut by the Tiber and its tributaries.

The Vatican

Located on a low extension of the Janiculum Hill that extends northeastward into the floodplain of the Tiber, the site of the Vatican has been occupied since Etruscan times. This low ridge is about 60 meters (197 feet) high and consists of Pliocene age claystones and sandstones, which are overlain by Pleistocene age stream deposits and tuffs from the Alban Hills volcanic field. Thick alluvial deposits wrap around the toe of the hill.

The Circus of Gaius and Nero, the first large structure to occupy the site, was tucked into the hillside and extended out over the alluvial plain. In A.D. 64, when the city of Rome burned, Peter and Paul were reputedly carrying out their apostleships. At the beginning of the 1st century A.D., Peter's tomb, adjacent to what was Nero's circus, was at the center of a growing necropolis. Around A.D. 320, Emperor Con-

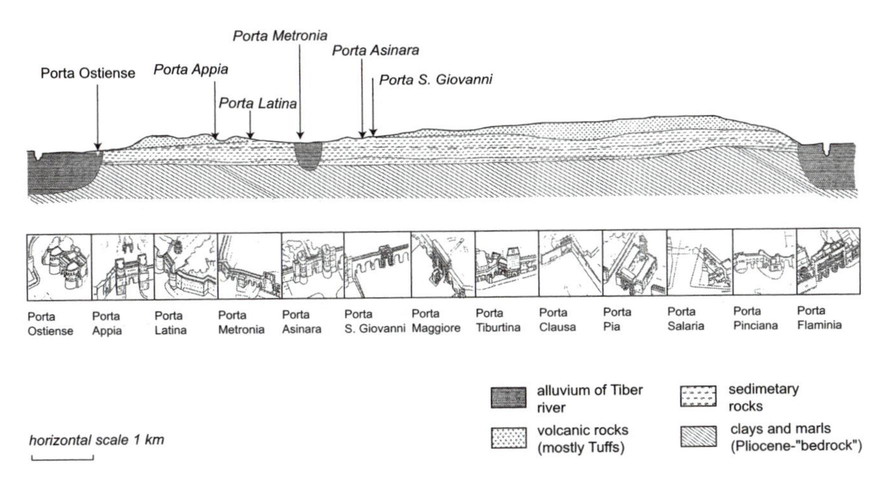

Geologic cross section below the Aurelian Wall. Gates most heavily damaged by earthquakes are those located on alluvium rather than rock.

stantine ordered the first basilica dedicated to Saint Peter to be constructed along the east-west base of Nero's Circus. Constantine's basilica survived for more than 1,000 years but by 1452 was in desperate need of repairs. Rather than rebuilding the old basilica, Nicholas V wanted a new church, but the old basilica was not torn down for another 50 years. Julius II hired Bramante to destroy the old church and to design the new one, for which construction took 120 years.

During construction of the present Saint Peter's Basilica, designers proposed that two campaniles (bell towers) flank the entrance. Work on the first bell tower began in 1612 and continued until 1641, but a sinking foundation required a scaled-down version of the original design, and the plan was in trouble. Bernini and others made an effort to save the campanile, but Pope Innocent X ordered its demolition. The problem, probably exacerbated by earthquakes, was that the towers were located on the alluvial plain and not on rock. Saint Peter's Basilica itself is located on rock and doesn't suffer the consequences of an unstable foundation. The contact of rock with alluvium runs across the front of the basilica, more or less at the head of the piazza (plaza). The piazza, flanked by 284 travertine columns, is an ideal construction on alluvium; it's a monumental space—exactly what was needed to set off the basilica's facade and to host the large crowds that attend ceremonies there.

More Clues from Conservation and Restoration of Historic Buildings and Monuments

From the beginnings to A.D. 2000, monuments and important buildings were often restored following earthquakes, floods, and significant political events such as Alaric's sack of Rome in A.D. 410, as well as during preparations for Jubilee years. The amount of money spent on restoration is useful as an indicator of actual earthquake damage following documented events. Engineers can use such seemingly unrelated historical detail, information from drilling, and models of earthquake effects to evaluate the risk to new structures planned for Rome. Geologists and engineers from the National Seismic Service and the Ministry of Public Works have evaluated seismic risk for every building in the historical center of Rome, as well as for the city's infrastructure, including gas and water distribution systems.

For the Jubilee year 2000, Rome undertook the mammoth task of not only cleaning most of its major monuments (and you are fortunate to see the glorious results today) but also retrofitting them to withstand future earthquakes. The magnitude of the effort required the combined financial support of resources available to architectural historians and archeologists and those from the Ministry for Civil Protection.

Most of the work to evaluate Rome's earthquake vulnerability has focused on the historical center. The effort now must be extended into the suburbs, where scant attention has been given to the geologic foundations beneath new apartment complexes and industrial and business centers. The most growth, which has occurred since the great earthquake of 1915, has taken place in areas that were open countryside at that time, and therefore we have limited data to help us predict the potential risks of earthquake damage in these newly built-up locations.

In Rome, geoscientists and government groups are working together to find ways to help citizens avoid earthquake-related disasters. Urban administrators in other parts of the world could learn from the Roman example: look at possible earthquake sources (the geologic foundations

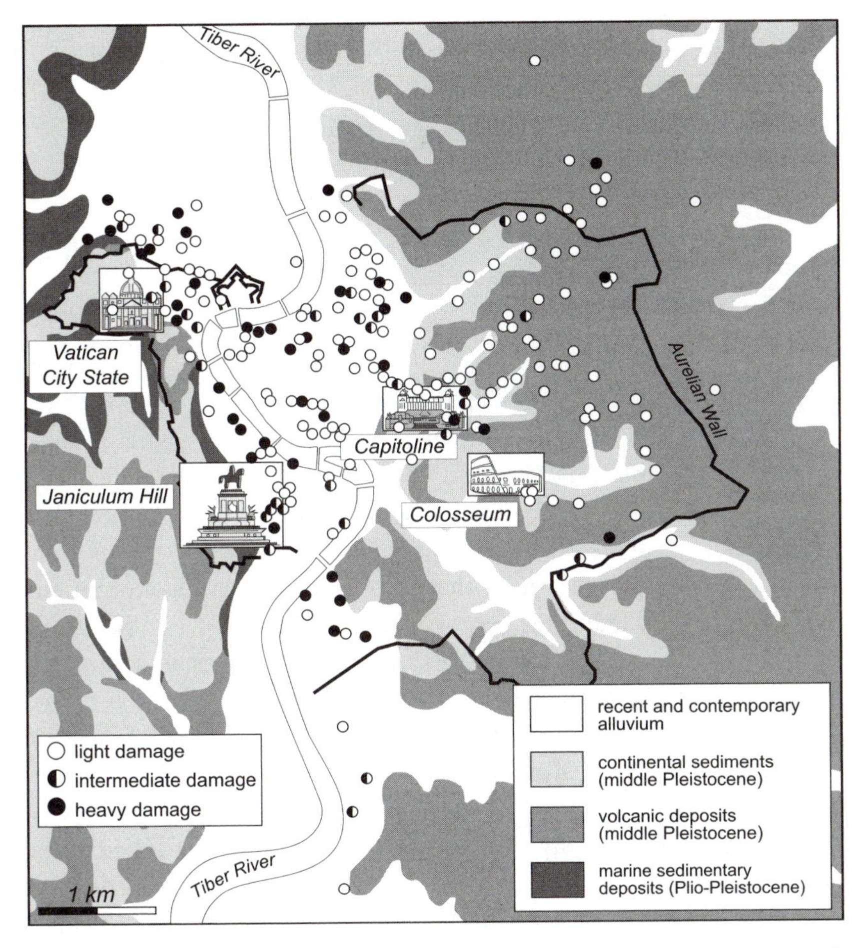

This map showing the basic geology of the historical center of Rome indicates those monuments that have required repair and restoration after earthquakes. The most heavily damaged areas are those on the Tiber River floodplain, where the sites are underlain by thick deposits of river sands and gravels.

upon which their cities have been constructed) and examine records of what has occurred in the past. No one has yet predicted an earthquake, but it is possible to predict structural damage by an earthquake before it happens and find ways to mitigate the dangers.

CHAPTER **7**

The Western Heights

JANICULUM, VATICAN, AND MONTE MARIO

RISING ABOVE the Trastevere and Prati neighborhoods, the Janiculum Hill, Vatican Hill, and Monte Mario are sometimes referred to as the "balcony of Rome." At the crest of the Janiculum Hill is the Piazzale Giuseppe Garibaldi, where a monument honors the father of modern Italy; this is arguably the very best viewpoint from which to contemplate Rome (other than from the window of an airplane, of course). Standing here, you can test yourself by identifying Rome's historic bluffs. It's easiest to see the geologic structure and the urban landscape on a clear day in the early morning or late afternoon. The tuff plateau extending out from the Alban Hills volcanic field is on your right. Directly in front of you, beyond the city, are the rugged Apennines, represented by the Lucretili, Tiburtini, and Prenestini mountains. Look carefully, and you can see the smooth surface of the ignimbrite plateau, incised near the Tiber by ancient streams that left the subtle relief characteristic of the famous seven hills.

Although the Janiculum Hill was an important part of Rome even in antiquity, it remained an outlying district of residences, sanctuaries, and light industry. Today the Janiculum is home to parks, the botanical garden, hospitals, and quiet residential neighborhoods. At its elevation of about 63 meters (206 feet) above the Tiber floodplain, the hill is cooled by summer breezes—one of the reasons it was the site of elegant upper-class villas before being developed as a residential suburb.

Reaching out northeast from the western heights, the Vatican Hill houses the offices and gardens of Vatican City. Saint Peter's Basilica is tucked between the western hills and the Tiber floodplain; its piazza is located on the riverine flats. Monte Mario, one of the highest and most isolated of the western hills at 124 meters (407 feet) above the Tiber, has been home to papal palaces, "country" villas, and Rome's astronomical observatory.

Collectively, this "balcony of Rome" overlooks the right (west) bank of the Tiber and holds the river to its north-south course before it turns westward toward the sea near Fiumicino Airport, thus separating Rome and the sea. The hills are composed of mudstones, claystones, and sands left by intermittent incursions of the sea over the last 2 million years. The many marine fossils support this interpretation of brinier times in Rome. During explosive eruptions in the Sabatini volcanic field north of Rome, upthrown ridges of marine sedimentary rocks were coated with pyroclastic flow deposits that swept down toward the Tiber. Rectangular, their long axes trending north-south, the western hills are an expression of "normal" faulting ("normal" refers to vertical offset along a fault). One of these north-south faults crosses Saint Peter's piazza and goes along the western margin of the Janiculum Hill; it may be responsible for the ramrod-straight valley that bisects the Janiculum and is now followed by the Viale dei Quattro Venti.

• • •

As you walk along the east side of the Janiculum Hill, you might like to take a quiet break in the Villa Sciarra, one of Rome's more attractive parks. Located above the Via Dandolo and the Trastevere neighborhood, and south of the Aurelian Gate in the city walls, the Villa Sciarra was home of the German Academy of Rome. The park lies along the eastern margin of an upfaulted block of marine sedimentary rocks, which are now overlain by 5 to 10 meters (16 to 33 feet) of man-made debris that have increased the slope's instability. Some of the park's grass- and tree-covered slopes are steep; gravity is pulling them down slowly, bending trees and accumulating rumpled clays and sands held back by brick walls above the Via Dandolo. This is not a problem of immediate danger but one of slow destruction, especially if Rome had a period of prolonged, heavy rainfall.

Summits like these along the eastern edge of the Janiculum provide beautiful, unencumbered views of the city, and it's easy to understand the temptation to build villas and apartments here. However, there is always the possibility that such perfect overlooks will end up at the bottom of the hill. Rome's situation isn't as severe as that of

On the slopes above the lower entrance to the Villa Sciarra park from the Via Dandolo, you can see evidence of the gradual slumping movement in the curved trunks of trees.

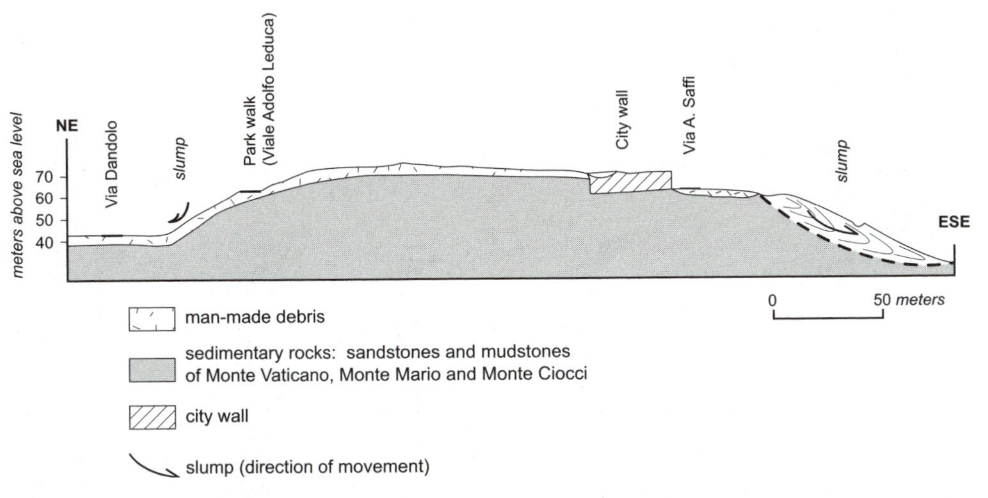

This geologic cross section through the Villa Sciarra park, from the Via Dandolo to the Via A. Saffi, along the edge of a hill above Trastevere, gives you a clear idea of the way the hillside has moved downward.

Los Angeles, California, where the combination of poorly consolidated sedimentary rocks, hillsides, and great views is commonplace, and dozens of homes end up sliding downslope after each heavy rain. Nevertheless, Rome's city planners should take heed of the potential for annual disasters.

Roman Walls

As you walk on the Via A. Saffi below the southern edge of the Villa Sciarra, you'll see another instance of slumping, which is tearing apart a section of the city wall, allowing it to gradually creep across the street. Rebuilding the wall and covering bare outcrops with concrete has done little more than slow the structure's inevitable downward progress.

This was once a section of Rome's defensive walls—one of many constructed in what was frequently a vain attempt to protect the city from invaders. The walls begun by Emperor Aurelian (A.D. 270–75) and completed by Emperor Probus (A.D. 276–82) were augmented by Maxentius (A.D. 306–12) and Honorius (A.D. 393–423) and then repaired and restored by the papacy again and again throughout time.

These brick arches across the Via Fratelli Bonnet in Monteverde are part of Rome's Aurelian Wall.

The last successful use of Rome's Aurelian Walls was as recent as 1849, when Garibaldi defended Rome from the French. For an excellent background on the history and engineering of Rome's defensive walls, visit the Museo delle Mura (Museum of the Wall) at Porta San Sebastiano; it is one of the city's most interesting (and overlooked) museums.

Main lines of defense, the 18.3 kilometers (11 miles) of walls enclosed the city of the time. The original Aurelian Wall included 381 towers, spaced every 30 meters (98 feet), and eighteen gates located astride the major consular roads; it also incorporated existing monuments, including the Pyramid of Caius Cestius near Porta Ostiense, also called Porta San Paolo. Along the crest of the Janiculum Hill, the main gate in the wall was the Porta Aurelia, also called Porta San Pancrazio, which protected the Via Aurelia at the edge of the city. Pope Leo IV extended the wall to guard the Vatican and Castel Sant'Angelo after the Saracen invasion of A.D. 846. More walls were added in the 17th century to extend protection to other parts of the Janiculum Hill.

In this modern clay quarry near Rome, claystones are visible as the gray bands near the middle and at the base of the sequence of Pliocene marine sedimentary rocks.

These walls have cores that are mostly rubble (stone, brick, and soil), but they are faced with brick and stone. The brick for construction and repairs was made from clay quarried in 2-million-year-old gray claystones, which are similar to the rocks underlying the western heights of Rome.

Brick has been commonly used throughout Rome. Until the 2nd century A.D., walls were often faced with small stone blocks arranged in a haphazard pattern (*opus incertum*) or later in a diamond pattern (*opus reticulatum*). Brick production was common in the southern Italian Peninsula and gradually spread north during the 1st century B.C.

The great Roman fire of A.D. 64 stimulated brick production as builders sought less flammable building materials. Wealthier people found it profitable to establish their own kilns for making brick and tile from the clay deposits on their estates. Business in brick and clay tile was remarkably profitable, and in the 3rd century it became an imperial monopoly.

The fact that most of Rome's defensive walls are still standing is testimony to good engineering, good building materials, good craftsmanship, and good maintenance. As we walk around the Janiculum, we pass the massively built Porta San Pancrazio, where the Via Aurelia entered Rome from the Tyrrhenian coast and cities in the north, then crossed the Tiber via the Pons Aemilius. In 1598 the bridge was destroyed by flooding, but remnants (now called the Ponte Rotto, or "Rotten Bridge") still stand south of Tiber Island.

Observing the traffic on the Via Aurelia, we are reminded that Rome's roads were as remarkably engineered as its walls. The ancient Via Aurelia, now used by commuters as a narrow but direct route into and out of the city, may be mostly covered by asphalt, but beneath it lies the original roadbed created by Imperial engineers.

Roman Roads

The Roman Empire's success was based in part on rapid communications and the ability to promptly move armies to areas of conflict—basic concepts that continue to govern modern nation-states. A series of long-distance roads linked Rome with its colonies and much of Europe (hence, the phrase "All roads lead to Rome").

The first engineered Roman road was the Appian Way (Appia Antica); construction began in 312 B.C., with funding from the censor Appius Claudius Caecus, to connect Rome with Benevento. The road was extended in 190 B.C. to the ports of Brundisium (Brindisi) on the Adriatic Sea and Tarentum (Taranto) on the Ionian Sea. The censor Gaius Flaminius began a second major phase of road building in 220 B.C. to connect Rome with Rimini and to establish a line of communications across the Apennines. The Via Flaminia linked strategically important towns and allowed Rome to extend its influence across the Italian Peninsula.

As the road network spread across Italy and then around the Mediterranean, engineers began straightening highways by employing bridges, open cuts, and tunnels. For example, in 174 B.C. the censors Flaccus and Albinus let contracts for the construction of many bridges and road resurfacing near Rome. Paving a road wasn't cheap even then; Diodorus Siculus writes that Appius bankrupted the state treasury with his road-building projects.

Throughout Europe, Roman roads still exist several thousand years after they were built. Many have been buried by subsequent road projects, but preserved Roman roadways can be seen in and around Rome and throughout Italy. The roads that we can see today are preserved because of their excellent base and a surface of finely crystalline basaltic lava blocks. Although suburban or urban roads were often surfaced with lava blocks, farther from the cities, Roman roads were surfaced with gravel.

Constructing a stable roadway in any era is a complex process: planners must understand both the engineering and the underlying geologic materials. Roman engineering technology appears to have been particularly advanced for its time, as is evident from the remnants of roads still intact. Chevalier's book on Roman roads includes the following checklist for a Roman civil engineer:

- Follow established paths; trace furrows along the sides of the future road.
- Excavate a trench of 1 to 1.5 meters (3.3 to 5 feet).
- Refill the trench with layers of foundation material, beginning with flat stones covered by several layers of concrete, followed by sand and gravel.
- Place curbs.
- Top with gravel or paving stones, making certain that the road slopes from its center to the sides for good drainage.
- If paving stones are used, use wedge-shaped stones (*gomphi*) to jam between the curbs and paving stones.

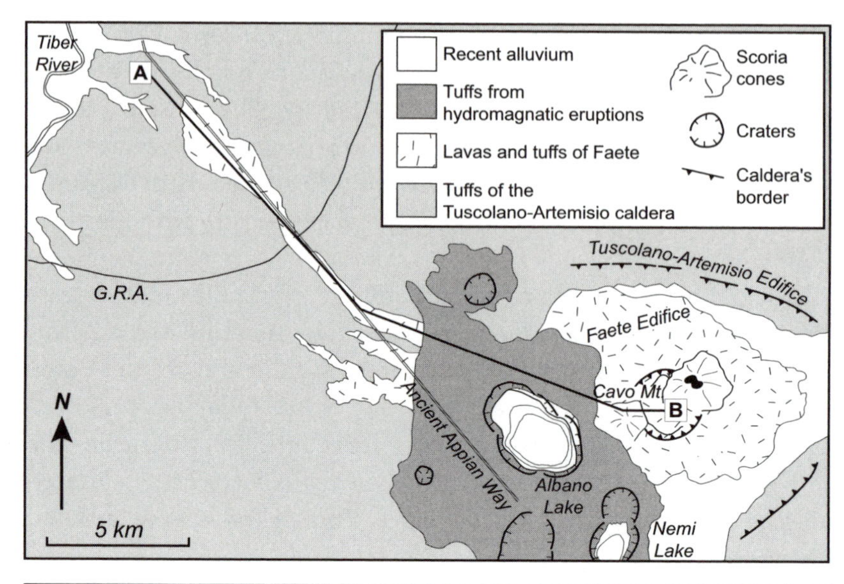

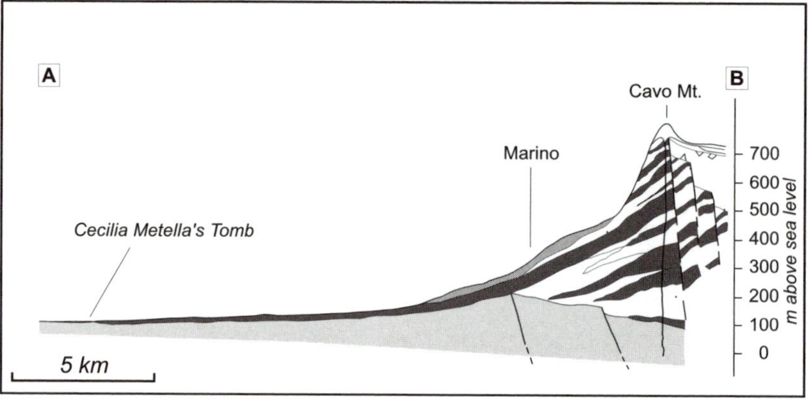

This geologic map and cross section show the northwestern Alban Hills volcanic field. The cross section (at a 10× vertical exaggeration) follows the Capo di Bove lava flow, which erupted from the Faete cone in the central part of the volcanic field and flowed northwest toward the ancestral Tiber River. The lava flow is named for the "head of the ox," which adorns the Tomb of Cecilia Metella, near the head of the Appian Way (near "A" on the map). The Appian Way followed the stable flow surface and was paved with blocks of lava quarried from the same flow. "GRA" is the Gran Raccordo Annulare, Rome's ring road.

One of the unique pleasures of visiting modern Rome is that you can still actually walk a archetypal Roman road: the Appian Way. This road follows the surface of a straight, narrow lava flow from a volcano in the central Alban Hills (the Capo di Bove lava). The flow's surface provided a firm foundation for the road as it rose into the hills southeast of Rome; it was also a source of the stone needed to construct stable roadbeds here and elsewhere. Romans preferred basaltic lava to all other paving stones. A road paved with basalt blocks was durable, as we can certainly see in remnants that have lasted for several thousand years. The large blocks used for road surfacing, called *basoli*, are the roughly worn surfaces that you often stumble across when visiting archeological sites. Later road surfacing, from the Renaissance to modern times, also used basalt blocks (especially in the cities), but they were much smaller and easier to handle. Even smaller, truncated pyramidal blocks used today are *sanpietrini*, named from their use in the plaza of Saint Peter's Basilica. The *sanpietrini* can be laid down to create designs in the road surface, thus allowing the builder to satisfy an appetite for art. You'll see piles of fist-sized *sanpietrini* throughout Rome where modern streets are being resurfaced. Many of these stones are undoubtedly recycled from earlier street work, but some have been imported from other countries; it's fascinating to speculate about the history they have witnessed.

Basaltic lavas are commonly quarried for road construction throughout the world and certainly throughout central and southern Italy. Nearest to Rome is the Capo di Bove lava flow that was erupted about 280,000 years ago from the Faete volcano in the Alban Hills. This flow followed an ancient valley toward the Tiber for about 20 kilometers (12 miles) and into what is now Rome. The Capo di Bove was named for the ox head perched on the Tomb of Caecilia Metella at the end of the flow. This basaltic lava was an excellent source of blocks for paving Roman roads, and some of the quarries are still visible along the northern edges of the flow.

Below the *basoli* or *sanpietrini*, Roman practice required a complex base of stone, sand, and gravel. In contemporary Rome, all types of construction produce a continuous demand for sand and gravel; every day, countless truckloads of sand and gravel are delivered in the city along the narrow, busy streets.

As you look at the surface of the ancient Via Labicana at Porta Maggiore, you can see the tracks from several millennia of traffic. The road's large lava blocks, quarried in lava flows near the city, are *basoli*; smaller blocks, *sanpietrini*, also used to surface the streets of Rome, were extracted and dressed by hand in nearby quarries. Interestingly, modern repairs are being made with lava blocks imported from China.

Granular Necessity—Sand and Gravel

Construction in all modern cities depends heavily on sand and gravel: as aggregates in concrete and asphalt, for roadbeds underlying the asphalt (or, in the case of Rome, under stone blocks), and as fill in utility trenches. A city the size of Rome uses about 18 million tons of sand and gravel each year. Thus, if sand and gravel quarries are, on average, 10 meters (32.8 feet) deep, the area annually affected by operations is 1.3 square kilometers, or 321 acres. The land encompassed by the quarries is also eliminated for any future agricultural use because the topsoil and underlying sediment have been removed; this environmental consequence is one of the penalties of urban growth.

What is the source for the millions of tons of sand and gravel that cities need for construction and repair? Typically, it is ancient river sediment or sand dunes. In the vicinity of Rome, sands and gravels were deposited during the Pleistocene (over the last 1.7 million years) by the ancient Tiber and by the advancing and retreating shorelines of the Mediterranean that followed the rise and fall of sea level. When sea level dropped (to as low as 120 meters [394 feet] below the present sea surface), the Tiber and other rivers feeding into the Tyrrhenian Sea cut their channels correspondingly deeper. Where the Tiber passed through what was to become Rome, it left a 50-meter-deep (164-foot) channel. As sea level rose, so did river level, subsequently filling the channel with sand and gravel. Although the Tiber's old channel in the midst of Rome could be a convenient and rich source of aggregates, you don't need to worry about developers excavating 50-meter-deep quarries in front of the Castel Sant'Angelo: the quarries would be quickly flooded by groundwater below a depth of 5 meters—and it would be culturally disastrous!

Ancient sand dunes that rimmed former Mediterranean shorelines are another source of aggregates. These coastal dunes, much like those that are active along the shores today between the mouth of the Tiber and Monte Circeo, are part of the sedimentary sequence left as the coastline moved back and forth while the sea rose and fell.

The abundant sands and gravels left by the ancient Tiber and coastal sand dunes during middle Pleistocene times (700,000 to 125,000 years ago) are quite accessible in Ponte Galeria, between Rome and the Fiumicino Airport. The quarries here were major aggregate sources for Rome but are now flooded because sediments were removed to a level below the groundwater table. Although the quarrying has caused severe environmental damage, the area now has a certain appeal for geologists—the sedimentary sequence is exposed for study and not hidden by soil or urban sprawl. The Ponte Galeria Formation contains a record of geologic history, documenting the changing landscape, climate, and geologic processes during interglacial and glacial periods. In these quarries, we see exposures of river conglomerates (consolidated gravel deposits) at the base, marine gravels and sands in the middle, and fine-grained lake sediments interbedded with tuffs from the Alban Hills volcanic field at the top. This sedimentary sequence is rich in fossils, including mollusks and vertebrates from above and below sea level. With their rich record of geologic history, some of the abandoned quarries have been chosen as "geotopes," or geologic parks, that can be used for research and teaching.

> Before aggregates are used in a construction job, they are sieved at the quarry and separated into sizes ranging from fine sand to coarse gravel. The energy involved in the natural processes of river transport and wind transport has already done a good job of separating the aggregates. By using these ancient deposits, quarry owners find that nature has already done much of their work for them.

The abundant supplies of sand and gravel would have not been sufficient to sustain Rome's continuous urban development over several millennia if the city had lacked the third major component of concrete: water. As we discuss the Celian Hill in the next chapter, you'll see how the accessibility and quality of water in the area have played their role in Roman history.

CHAPTER 8

The Celian (Celio) Hill

ALTHOUGH it is obviously a prominent plateau and important in Roman history, the Celian Hill is off the typical tourist's beaten track. Located across the Via Celio Vibenna from the Colosseum, it provides a pleasant escape from the chaos that swirls around that great arena. The northern edge of the hill retains remnants of the Temple of Claudius and the Churches of San Gregorio Magno and Santi Giovanni e Paolo, as well as several very pleasant parks. Take one of Rome's colorful trams that follow the line from the Porta Maggiore to Piazzale Ostiense, and then rattle up and across the northern edge of the Celian Hill above the Colosseum. Get off at the Parco del Celio stop and spend a few minutes enjoying an overview of the Colosseum from the base of the Temple of Claudius.

Like the other six hills, the Celio was created by erosion of the pyroclastic flow deposits (tuffs) left after volcanic eruptions in the Alban Hills. This 700-by-400-meter (69-acre) hill is linked to a less eroded part of the tuff plateau that underlies San Giovanni in Laterano. If you walk downhill from the Parco del Celio tram stop, you'll be stepping into the stream valley separating the Palatine and Celian hills, a route now traced by the broad Via di San Gregorio. At the southwestern corner at the bottom of the Celian, stone steps lead upward to the Church of San Gregorio Magno, founded in A.D. 575 by Saint Gregory the Great. As you continue uphill via the Clivo di Scauro, you'll reach Santi Giovanni e Paolo, a 4th-century church built adjacent to the Temple of Claudius. During work on one of many additions and renovations, the 13th-century bell tower was built upon the platform of the Temple of Claudius; you can see this historical layering most easily in the forecourt of the church. While you are here, look at the Claudian construction: these immense blocks are travertine, one of the architectural stones most commonly used in Rome since the days of the Republic.

The Ponte Cestio, the bridge from the Tiber Island to the right bank of the Tiber, was built in 46 B.C. and rebuilt in the 19th century with the same stone. The large white blocks, with the pockmarked texture, are travertine, a banded calcium carbonate spring deposit. The Roman region has many travertine deposits, but the best are east of Rome at Bagni di Tivoli. Romans have used and continue to use large volumes of travertine for construction and works of art.

"Icing on the Cake": Travertine

Travertine is a light-colored (mostly white), massive, finely crystalline form of calcium carbonate; its chemical composition is similar to that of limestone and marble, but it has a very different texture. You'll find this elegant stone throughout the city in Renaissance sculptures, blocks in ancient Roman buildings, stone facings in your hotel, and curbstones along the street. The word *travertine* was derived from the Latin *Tiburtinus* (from Tivoli), reflecting the importance of this rock to Imperial Rome.

Rainwater that falls on the Apennines is slightly acidic because of dissolved atmospheric carbon dioxide. Circulating through cracks in the limestone that is exposed in the high mountains, these waters dissolve some of the limestone, leaving behind cave networks and open fractures. If the groundwater circulates deeply enough, it is heated and mixes with geothermal fluids, which add more calcium and bicarbonate. When the mineral springs issue from faults, the waters flow across broad areas to form widespread travertine deposits.

How Travertine Is Formed

This beautiful stone is the product of crustal motions (faulting), the rise of magma and gases (volcanism), and circulation of groundwater.

Travertine is formed by the chemical precipitation of calcium carbonate from solution in the groundwater. Thin accumulations of this chemical sediment leave distinct layers—their thickness depending on the spring's flow rate and water composition. As you examine a piece of travertine, you can see that variations in flow rate and composition have produced a "banded" rock, in which layers of airborne dust and debris have accentuated the sequences. This attractive layering, plus the ease of quarrying, has made banded travertine very desirable for both construction and sculpture.

The rates for travertine deposition vary from 0.4 millimeters (0.02 inches) per year (producing a very uniform rock) to as much as 21 centimeters (8.4 inches) per year (generating a coarser rock with many holes). The faster depositional rates have actually encased beer cans near springs where people carelessly left their picnic trash. Such environmentally insensitive behavior has had one positive effect in that it has provided a time marker—we know that "easy open" cans were introduced in 1960! You can imagine that, if one has enough patience, travertine could be a renewable resource, although perhaps not quite renewable enough, given the modern demand for this stone.

The Roman travertine is found mostly in the Acque Albule basin, located near Tivoli (ancient Tibur) at the foot of the Tiburtini-Cornicolani Mountains. This basin has been gradually pulled apart by tectonic activity associated with north-south-oriented faults and has subsided as much as 200 meters (656 feet) over the last 400,000 years. The Acque Albule travertines, covering 30 square kilometers (11.5 square miles) and ranging from several meters (6 feet) to as much as 85 meters (280 feet) thick, were deposited in shallow lakes by springs along the basin edges over the past 165,000 years. Claudio Faccenna and his colleagues at the University of Roma Tre have been able to establish a link between times of increased deposition and periods of faulting, explosive volcanic eruptions, and greater hydrothermal activity in the Alban Hills: faulting and volcanism fractured the aquifers, adding gases such as hydrogen sulfide and carbon dioxide to the springs.

Although the travertine deposits near Tivoli have been quarried since ancient Roman times, the stone was especially popular during the Italian Renaissance and the Fascist periods. Early Romans began using travertine but its use was limited because the stone was difficult to cut with the tools of the time. Later, during Imperial times, improved technology allowed wider use in public buildings such as the Colosseum. Now vast industrial quarries use the latest techniques to produce mostly thin slabs to be used as building facings.

• • •

You'll see many examples of fine travertine in other monuments on the Celian Hill, such as the Arch of Dolabella.

As you continue your trek along the Clivo di Scauro, the street becomes the Via di San Paolo della Croce. At this point, you have several choices:

- You can continue downhill to the Via Claudia at the Arch of Dolabella, which is yet another travertine construction built in A.D. 10 by consuls Gaius Junius Silanus and Cornelius Dolabella. This arch originally may have been a gateway in the Servian Wall, but it was later incorporated into an extension of the Claudian aqueduct that supplied water to the Palatine Hill.
- Or, if you turn south into the park of Villa Celimontana, you can rest or have a quiet picnic. The Dukes of Mattei purchased what was a vineyard in the mid-16th century and turned it into this formal garden that is now owned by the city of Rome. From slopes on the south side of the park there are some remarkable views of

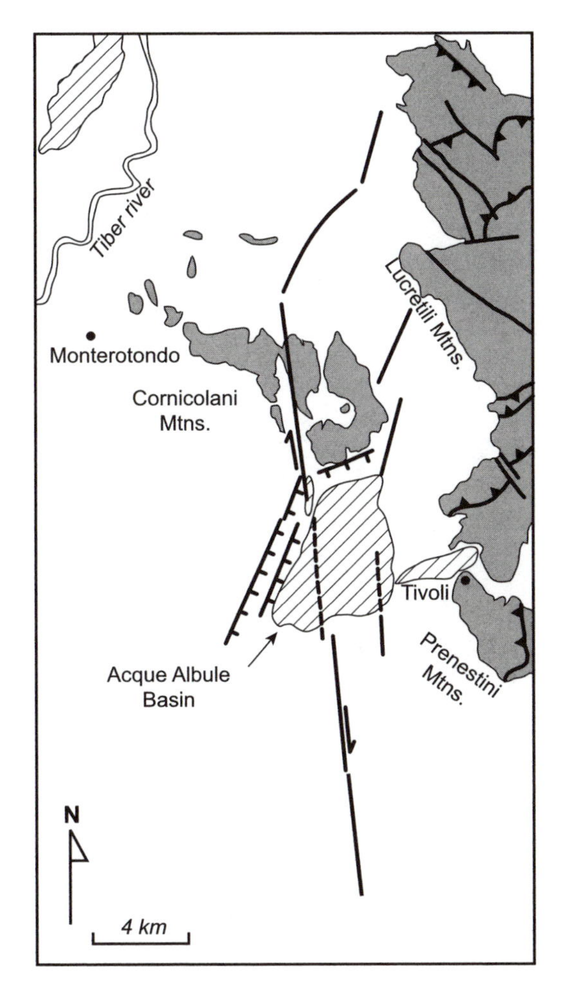

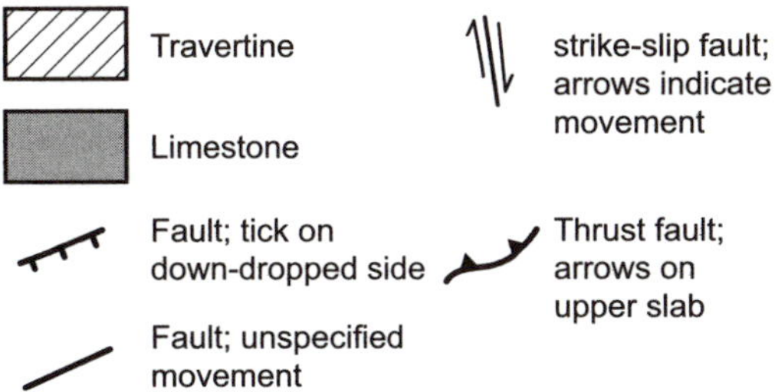

At the center of this geologic map of the Monte Cornicolani–Tivoli area east of Rome is the Acque Albule basin, which contains a travertine resource of approximately 2.55 cubic kilometers (0.61 cubic miles). Travertine is deposited by mineral springs rising to the surface along faults at the base of the Apennines or close to the volcanic fields of Sabatini (northwest) and the Alban Hills (south).

This modern travertine quarry, Bagni di Tivoli, is located east of Rome near the town of Tivoli. The travertine here is of very high quality, displaying few of the pockmarks or cavities typical of most travertine.

the Baths of Caracalla. At one time Rome had many public and private baths; those of Emperor Caracalla (A.D. 211–17) are the most noteworthy in modern Rome.

The Baths of Caracalla

Fed by the aqueducts that brought water from the Apennines into Rome, the public baths are symbolic of the close relationship between the city's water resources and the success of the Empire. As it was transported through cement-lined channels, water was protected from the elements, evaporation, and diversion. But once it reached the city, who benefited, and how was the water distributed?

Vitruvius states that the water supply went to public basins and fountains, public baths, and private consumers. Any excess water went to the public or was used to flush the sewers. Historical records show that Frontinus made certain that enough reservoirs were constructed to maintain an emergency city water supply in the event of drought or fires.

The demand for water increased not only with a growing Roman population but also with the increasing popularity of public baths. Beginning in the 2nd century B.C., private baths were established for those who could pay an entrance fee. Baths evolved, and their popularity grew in concert with a basic interest in cleanliness and a comfortable life. Eventually, the baths were opened to everyone regardless of social status. Public facilities included hot baths, steam rooms, tepid rooms, and cold plunges. The baths grew to include athletic facilities, shops, libraries, places to eat and drink, theaters, and parks—forerunners of today's malls. The city built new aqueducts or increased the capacity of an existing aqueduct whenever a new bath was opened. By the end of the 4th century A.D., Rome had 11 large public baths (*thermae*), 965 smaller bathhouses, and 1,352 public fountains for a population of about a million people.

In their book on the Baths of Caracalla, Leonardo Lombardi and Angelo Corazza quote an epitaph from the tomb of a citizen of the Roman Imperial state: "Balnea vina, Venus corrumpunt corpora nostra sed vitam faciunt" (or "Baths, wines, and Venus seduce our bodies, but give life"). Among the best-preserved *thermae* in Rome, the Baths of Caracalla opened in A.D. 217 under the regime of Caracalla and were the most beautiful of the time. A new branch of the Aqua Julia was added to compensate for the increased water use by an appreciative citizenry.

The baths required sophisticated hydraulic systems, complete with pressure controls, filtration or settling systems, and storage tanks, as well as concrete tanks for heating, and waste disposal into a sewer system. The first requirement, however, was to find water sources and then to transport the water into the city.

Water—And Enough of It

Water, not wine, is the lifeblood of Rome. During your visit, you'll note that fresh water is evident everywhere: in the fountains, in the aqueducts, and in its ceaseless flow from street taps. Water issues from the mouths of mythical creatures, from elegantly designed fontanelle, and from simple brass tubes. Potable water flows constantly and provides refreshment for walkers, runners, shoppers, workers, housekeepers, and beloved pets. At any spot in the city, you are rarely more than a short walk from a refreshing drink. At 450 liters (119 gallons) per capita per day, Rome's citizens have one of the most generous supplies of drinking water in the world. And it is pure even today.

Visitors to Rome have always been impressed by the constant flow of clean water. Johann Wolfgang von Goethe commented on its availability in 18th-century Rome, unusual at a time when water supplies for most European cities were so contaminated that their residents drank beer or cider:

> In the square in front of San Pietro in Montorio we paid our respects to the powerful current of the Acqua Paola, which flows in five streams through the gates and openings of a triumphal arch to fill an enormous basin. This great volume of water comes from beyond Lake Bracciano by an aqueduct twenty-five miles long, which was restored by Pope Paul V. This takes a peculiar zigzag course through the hills, satisfying the needs of various mills and factories on the way, until it empties into a wider channel in Trastevere.

> The lovers of architecture among us extolled the happy thought which had provided this water with a free, triumphal entry open to all. The columns, arches, cornices, and pediments reminded us of those sumptuous arches through which, in times past, returning conquerors used to enter in triumph. In this case, it is the most peaceful of benefactors that enters with a like power and is received with immediate gratitude and admiration for its long and strenuous march. Inscriptions inform the visitor that it is Providence and a beneficient pope of the Borghese family who are making by proxy their stately and enternal entry.
>
> Someone who had recently arrived from the north objected to the arch and said it would have been better to let the waters emerge into the daylight in a natural manner over a pile of rugged rocks, but we pointed out to him that they were not natural but artificial, so that it was only fitting to hail their arrival in an artificial manner. (Johann Wolfgang von Goethe, *Italian Journey*, December 1787)

Not much has changed since Goethe and his friends visited the Acqua Paola. It is still flowing, still impressive, and can be found by taking a short walk up the Via di San Pancrazio from Trastevere or from the southern end of the park on the Janiculum Hill.

The water arrives in the city via the Aqueduct of Trajan (Aqua Traiana), which is visible along the northern border of the Villa Pamphili park, near the end of its 35-kilometer (22-mile) run downslope from springs near one of the volcanic lakes that flank the city. The aqueduct was damaged and fell into disuse during medieval times. At the beginning of the 16th century, Trastevere's increasing population needed clean water: residents had only the most rudimentary means to filter contaminated water from the Tiber. Pope Paul V Borghese (1605–21) proposed rebuilding the abandoned Aqueduct of Trajan to restore water resources that hadn't been available since the fall of Imperial Rome. Pope Innocent X Pamphili strengthened the same aqueduct in 1846, adding water from Lake Bracciano. Today, water flows through the aqueduct at 1,000 liters per second (15,852 gallons per minute), but the flow can be raised to 8,000 liters per second (127,000 gallons per minute) for emergency situations.

Rome is the only city of its size in the world that is chiefly supported by groundwater in a sustainable manner. *Sustainable* here means that Rome can continue to do so, unlike other groundwater-dependent

The Acqua Paola is a monumental fountain established by Pope Paul V to celebrate water coming into Trastevere via the restored Aqua Traiana. The aqueduct is still used today to bring water from springs near Lake Bracciano, a volcanic crater lake located in the Sabatini volcanic field northwest of Rome. The fountain overlooks Rome from the Janiculum Hill and, with a great view of the city, reinforces the historic link between water and the city's fortunes.

cities, such as Mexico City, that are "mining" the resource and will eventually deplete their aquifers. Collectively, 24,000 liters per second (380,448 gallons per minute) flow into Rome, or, in grander terms, 756 billion liters (200 billion gallons) each year—enough to fill a swimming pool 2 meters (6.5 feet) deep and 20 kilometers (12.5 miles) on each side. In contrast with some cities in the United States, such as Los Angeles, San Francisco, or New York, where water supplies are collected at the surface and transported hundreds of kilometers, most sources for Rome's water are collected within a straight-line radius of 40 kilometers (25 miles). In addition, water supplies for those cities originate mostly as surface runoff, so there are great losses from evaporation or transpiration (water used by plants); Rome's underground water supply loses little to the atmosphere or to forests and grasslands.

Where are the sources for Rome's abundant water? The Tiber and a few springs within the city itself are minor sources, and some of the water comes from lakes and springs in the volcanic fields that flank the city, but most of it comes from vast aquifers within the distant Apennine Mountains.

Beginnings: Water from Springs in the City

Springs have always held symbolic and historic importance and been associated with gods and fairies during the early development of many cultures. Springs are commonly thought to create a sense of well-being, especially if the product is mineral water and is easily marketable. Spas throughout the world have traditionally traded on these natural phenomena; entire towns have been built up around well-known springs. However, those within the historic center of Rome are not large enough to take care of much more than a village. Until the 4th century B.C., when the Romans initiated a system of aqueducts, the young city relied on springs and water wells dug along the edges of the Tiber as its only water sources; it supplemented this supply by building cisterns to collect rainwater. Most of the springs within Rome are buried by the debris of history, paved over, or simply forgotten.

Springs emerge where groundwater, seeping through relatively permeable rock (water-bearing, permeable deposits, or *aquifers*), encounters another rock unit with little or no permeability at or near the

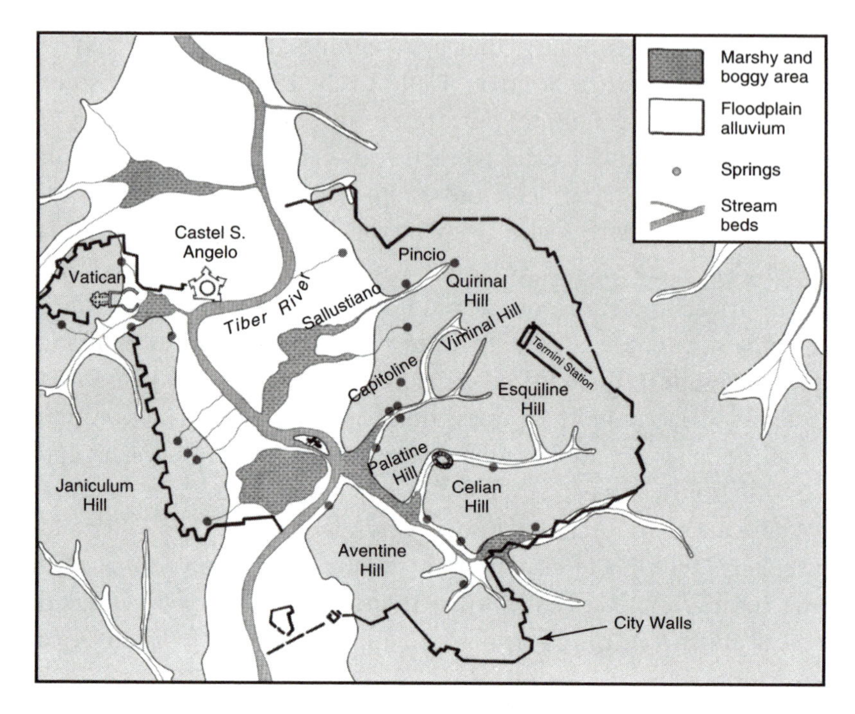

Springs in the historic center of Rome issued from water-saturated alluvial sediment or volcanic ash deposits that overlie poorly permeable marine claystones. (Adapted from Corazza and Lombardi 1995)

ground surface. Depending on the physical character of both geologic units, water can seep out evenly, forming a patch of wet, boggy ground, or from a point, where the water flows as if from a pipe. Many of the historic springs in Rome are located at the base of hills located on either side of the Tiber, especially above the left bank. There, waters seeping down through permeable sand and gravel deposits left by the ancient Tiber—or through moderately permeable volcanic rocks—come into contact with the older, impermeable clayey sedimentary rocks.

Some Historic Roman Springs

- Acqua Tulliana—below the Capitoline Hill; known by the Romans as Tullianum; its legendary fame is tied to its location in the cell of the jailed apostle Peter.

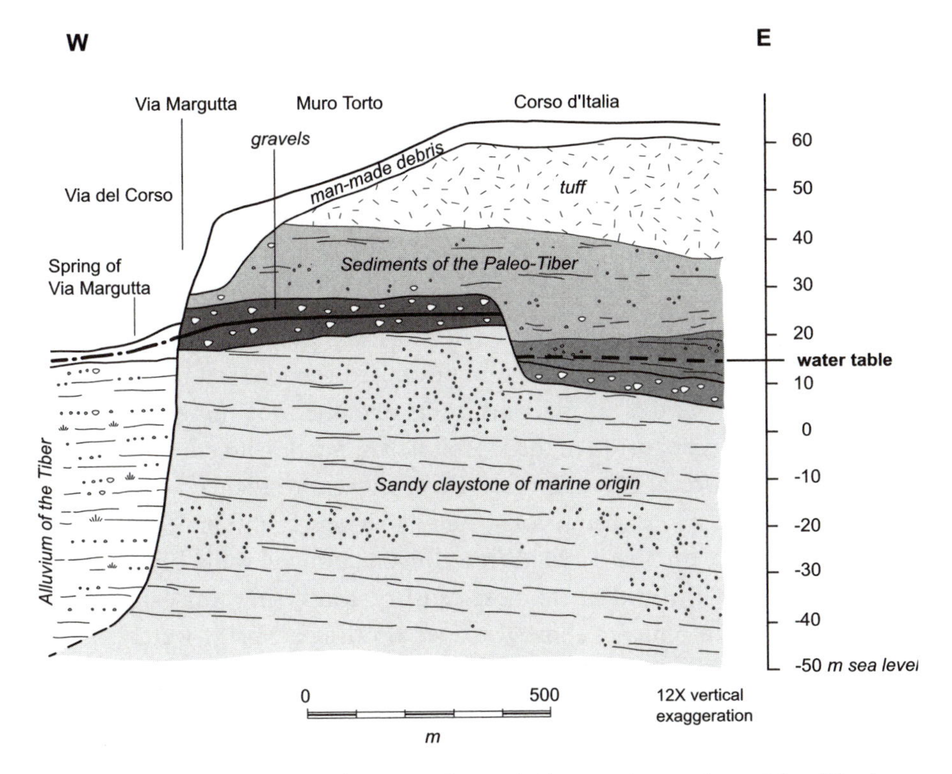

This geologic cross section cuts through the western edge of the Pincio, from the Villa Borghese gardens to the Via del Corso, near the Piazza del Popolo. The top of the water table sits above poorly permeable claystones, which are overlain by more permeable river sands and gravels and tuffs. The Spring of Via Margutta emerges (or once emerged, before being buried by the debris of the city) where the water table intersects the foot of the hill. (Adapted from Corazza and Lombardi 1995)

- Acque Lautole—in the Roman Forum; provided warm (37°C or 98.6°F) mineral water.
- Acqua di San Clemente—now located below the Church of San Clemente; emerges from a Roman water pipe.
- Fonte delle Camene (Latin Camenae)—the base of the Celian Hill, which, in ancient times, was rich with springs.
- Piscina Pubblica (Latin Publica)—near the Baths of Caracalla; accessed by a well that still stands in the baths.

- Acqua della Fontana degli Api/Acqua di Santa Maria delle Grazie—discovered in the 17th century, during development of the Belvedere Garden in the Vatican; marble fountain designed by G. Lorenzo Bernini; destroyed by construction of the Vatican railway station; fountain was moved to Piazza delle Vaschette.
- Acqua Damasiana—just outside of the Porta Cavalleggeri; discovered in 1756.
- Aqua Pia—on northern slopes of the Janiculum Hill near the Porta Cavallegeri; discovered during excavation of clay for a kiln.
- Acqua Lancisiana—several popes noted the medical benefits of water from this spring, and it was exploited for mineral water. In 1827, expansion of the Hospital of Santo Spirito cut off access to the spring. After a protest by Trastevere residents, Pope Pius XIII had an extension constructed to a new fountain near Porto Leonino on the Tiber. With construction of walls along the Tiber at the end of the next century, the spring was extended farther into niches above the river. Bottling of spring water continued, most recently under the labels "Acqua della Fontanella" and "Acqua della Barchetta."
- Acque Corsiniane—in the garden of the Palazzo Corsini (the Botanical Gardens at the base of the Janiculum Hill.
- Acque Sallustiane—in a valley between the Pincian and Quirinal hills; buried by the evolving city, but for a while one of these springs flowed in the basement of the Hotel Bristol Bernini near the Piazza Barberini.
- Acque di San Felice—in the garden of the Quirinal Palace, the San Felice fountain is fed by what probably was the ancient Cati Fons.
- Lupercale—exact location is not known but is somewhere along the base of the Palatine Hill; during Roman times, this spring was dedicated to Lupercus Faunus, the one who keeps wolves away, and the water was used for the Lupercalia feasts.

Water from the Tiber

Even during the youngest days of Rome, the Tiber was rarely used as a source of drinking water. Although it was once called Albula, for its clear, light color, the Tiber was polluted very early in Rome's history. Only in desperation, such as when aqueducts were destroyed by the Goths, did people drink from the Tiber. Today, Rome's river is alarmingly polluted, as are most streams in the Lazio region.

Water from the Volcanic Springs and Lakes

The volcanic highlands of the Alban Hills and the Sabatini volcanoes flanking Rome are the second most important source of water for the city. Rainfall soaks into the volcanoes' slopes—a catchment area of 5,100 square kilometers (about 2,000 square miles)—and recharges the aquifers below; the area provides a cumulative flow of surface water and groundwater of 45,000 liters per second (700,000 gallons per minute). Groundwater, slowly flowing downslope within permeable volcanic ash deposits or along fractures in lava flows, feeds springs that emerge along valleys that drain the volcanic fields. Springwater from the Alban Hills was (and still is) transported short distances (4.5 to 7.5 kilometers, or 2.8 to 4.7 miles) via the Tepula and Julian aqueducts, which feed into the Claudian aqueduct.

Bracciano and Albano crater lakes, the largest natural lakes in this region, are additional water sources for both Rome and regional agriculture. Trajan's aqueduct brings water from the springs located near Lake Bracciano in the Sabatini volcanic field and, since 1846, from the lake itself.

Water from the Apennines: Backbone of the Central Italian Peninsula

The greatest source of Rome's pure water is the structurally complex Apennine mountain chain. This range is made up of mostly sedimentary rocks that were deposited in ancient seas, subjected to elevated temperatures and pressures while deeply buried, consolidated, then thrust up to their present elevation, where they form the "backbone of Italy." Within the regions of Lazio and Abruzzo, the marine sedimentary rocks are mostly limestone (calcium carbonate) and dolomite (magnesian-calcium carbonates). With time, slightly acidic rainfall in-

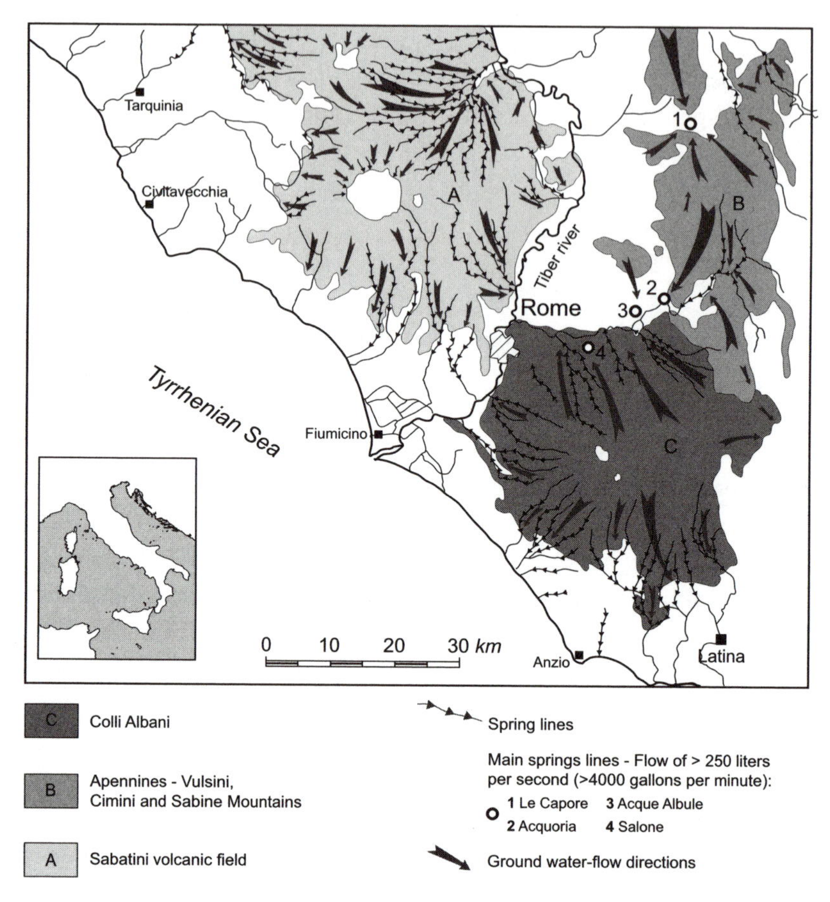

The general hydrologic framework of the Roman region includes the karstic aquifers of the central Apennines, as well as aquifers associated with the volcanic fields that flank Rome. (Adapted from Corazza and Lombardi 1995)

filtrates along fractures into these limestones, dissolves the carbonate, and forms cave networks. If a region has many open fissures, sinkholes, and caves, it is called a *karst* terrain. Such terrains are extremely unfavorable for agriculture because so much of the water goes underground and the soils are not very fertile. If you are a cave explorer or speleologist, however, karst will provide wonderful frontiers: caves featuring stalactites, stalagmites, and underground rivers.

The central Italian Apennines contain karst terrains over an area of about 8,000 square kilometers (3,100 square miles) and, if we calculate the infiltration rates and rainfall for the region, support a cumulative groundwater outflow of 220,000 liters per second (3.5 million gallons

This photograph was taken in the central Apennines, near the medieval town of Subiaco, where many of the springs that feed the Roman aqueducts issue from limestone karst.

per minute). Not all this water reaches the surface, but there is plenty to serve Rome's demand for potable water. Hydrologists P. Pace and P. Bono concluded that the largest flow through the ancient aqueducts from the karst springs of the Apennines was 18,433 Roman *quinariae* (764,454,000 liters, or 202 million gallons) per day, which was 76 percent of the water destined for Rome.

Copious springs emerge at the base of the Simbruini Mountains at elevations of 321 to 327 meters (1,053 to 1,073 feet) and flow into the Aniene River, which joins the Tiber in Rome. While driving up the Aniene River valley to the medieval town of Subiaco, you'll pass by these springs, which are mostly hidden from view. If you watch for them, however, you'll see pipes running from springhouses and wells into a modern aqueduct that is camouflaged with soil and grass.

Gathering the Water Before It Entered the Aqueducts

Once springs and shallow aquifers were identified, the water had to be "gathered" and fed to the head of an aqueduct. Not much about this concept has changed, although the ancient Romans obviously relied more on muscle rather than on the heavy equipment that we can command today. Vitruvius recommended digging multiple shafts or tunnels to test underground water supplies. After an aquifer was identified, tunnels were excavated below the water-saturated zone and were oriented to reach a common outlet. Streams captured at the source were also used to transmit water from catchment basins. Springs and streams flowed into settling tanks, where sediment could fall to the bottom before the water entered an aqueduct.

Roman Engineers, Public Health, and the Water Supply

Today almost everyone is concerned about clean water supplies, which is a reasonable obsession. Water sources are examined, reservoir waters are treated with chemicals to kill bacteria, and, in most industrial countries, supplies are regularly tested by water chemistry laboratories. In Italy today, the labels on bottled water not only give the nature of the spring and the chemical composition but often contain assurances by a university professor that the water is pure and safe.

Not having chemistry laboratories, how did the ancient Romans test their water supplies? They observed the people who used the springs—human guinea pigs. Vitruvius described the Roman criteria for identifying a safe water supply: it must be "visibly pure and clear, inaccessible to pollution, and must exhibit outflows free from moss and reeds. Study the general health of the local consumers with special reference to complexion, strength of bone, and clouded eyes. Samples should be tested in a good quality bronze container for corrosion, sparkle, pouring foreign bodies, and rapid boiling." The Romans understood that lead pipes, although they were used to some extent in the distribution systems, were unhealthy. We read that Vitruvius did not favor lead pipes because of lead poisoning and preferred almost any other material (stone, wood, or terra-cotta).

It's interesting to speculate that without the ability to identify clean water sources and the engineering expertise to build aqueducts for bringing the water to the city, Rome could never have developed into the seat of power it became.

Roman Aqueducts: The City's Lifeline

Before the 4th century B.C., Rome was a city with a respectable population of about 100,000 residents. The population stayed at this level, limited mainly by drinking water available from springs, wells, and the Tiber River. To bring water from the copious Apennine springs and the volcanic highlands, Roman engineers had to conceive and construct a sophisticated system of aqueducts (Latin *aquae ductus*, "conveyance of water").

The first of eleven aqueducts, the Aqua Appia, was completed in 312 B.C. and construction of more aqueducts continued intermittently over 600 years. With the construction of each new aqueduct presumably came an increase in population. We can estimate that in the 4th century B.C., Rome's water supply was about 1,500 liters per second (24,000 gallons per minute), and the estimated population may have been about 200,000. The Aqua Appia supplied 900 liters per second (14,000 gallons

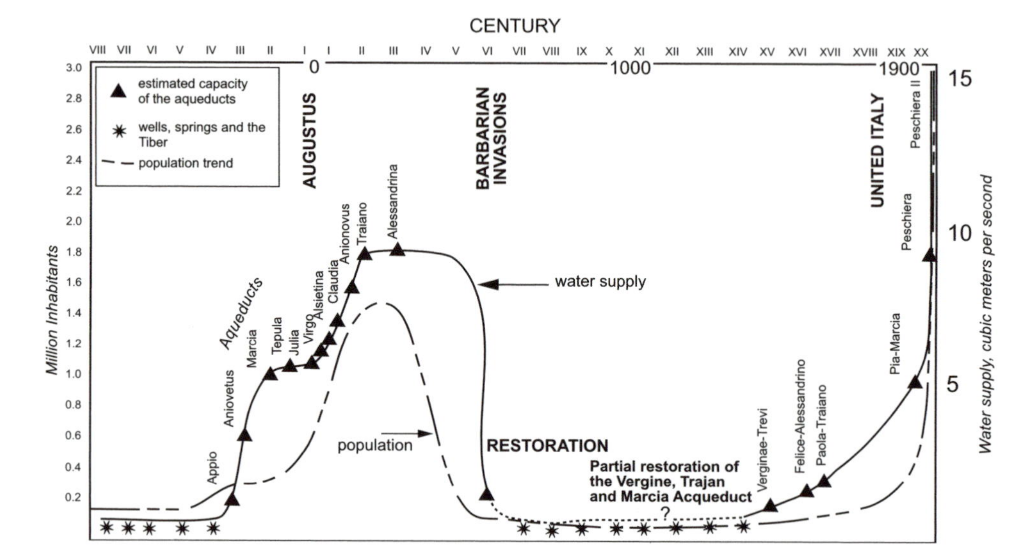

This diagram showing water supply and the population of Rome from the 8th century B.C. to the late 20th century A.D. reveals the strong link between a stable water supply and the health of a city. (Adapted from Funiciello and Rosa 1995)

per minute). After construction of the eleventh aqueduct, in the 3rd century A.D., water was being brought in to the city at 12,000 liters per second (about 186,666 gallons per minute), and the population had risen to more than 1 million people, making Rome the largest city in the Western world.

By the 6th century A.D., after several centuries of administrative neglect and the destruction of aqueducts during barbarian invasions, Rome became once again a small, rather miserable town, barely surviving for nearly a thousand years with a population of between 10,000 and 50,000. The aqueducts that were not destroyed were rarely maintained and eventually became clogged with carbonate deposits. Only the Virgo (Vergine) aqueduct maintained a constant flow rate. Much of the drinking water during this time came from the polluted Tiber, which increased the number of waterborne diseases. In addition to losing water sources, Rome's economy was affected by the loss of water mills along the aqueducts, which had been used to grind grain for the bread supply.

After medieval times, the aqueducts were gradually restored and expanded, and the population rose in parallel with a supply of fresh water. The 24,000 liters per second (380,448 gallons per minute) flowing into Rome today support a population of nearly 3.5 million.

All cities can learn from Rome's historical experience of the association between fresh water and a healthy urban environment. The correlation between accessible water and population is clear; it is difficult, however, to determine if water supply drove the population increase or if the need for water drove aqueduct construction.

Roman Aqueducts of the Classical Period

The aqueducts were an absolute necessity for the physical and economic health of Rome. Along with the roads and sewer systems found across the Empire, the aqueducts are the most famous examples of Roman engineering. Our admiration grows as we realize that many of their features are still regarded as state-of-the-art in water distribution systems.

To build concrete-lined, covered aqueduct channels required a detailed understanding of the terrain because water moved to the city entirely by gravity (obviously, no electrical pumps were available to these engineers). The aqueduct's slope, from spring and river sources in the mountains to the city below, dropped between 0.5 and 15 meters per 1,000 meters distance (2.4 to 81 feet per mile). Where aqueducts entered Rome, the water passed through "water castles" (water towers), from which it was distributed throughout the city. This system is basically the same as that used by ACEA, the water utility that supplies Rome today.

Aqueduct engineers had to maintain a uniform slope across the terrain while bridging valleys and tunneling through hills. This standard required surveying the terrain and constructing detailed maps. Surveying instruments of the time included the *groma* and *dioptra* (similar to the theodolite used today, but without a telescope), a water level, and a 6-meter-long (20-foot) field level that could be adjusted with plumb lines.

Our knowledge of the Roman aqueducts comes mainly from *De aquae ductu*, written by Sextus Julius Frontinus, the Imperial *curator aquarum* (water commissioner) during the reigns of Emperors Nerva and Trajan. Frontinus accumulated the knowledge for his treatise because he felt that, as the chief water commissioner, he should understand the workings of the water system. The main purpose of his office was to patrol, maintain, and repair any damage to the aqueducts—a charter much like that of any modern city's utilities department—and, like today, water leakage was a big problem. It's astounding to realize that in many modern cities, more than 50 percent of the water brought from distant sources is lost through broken or cracked channels and pipes (or through illegal taps into the line). Frontinus devised a simple means of determining both the amount of water going into the aqueducts and the water volume that reached Rome. When, in addition to identifying maintenance problems, his staff discovered numerous illegal taps into the lines, the taps were removed and the water thieves punished. Frontinus's goal of maintaining a clean supply of water for the citizens of Rome was successful through routine maintenance and enforcement.

The initial Roman aqueduct, the Aqua Appia, was built by the censors Appius Claudius Caecus and Caius Plautius in 312 B.C. This 15-kilometer-long (9.3-mile) feature was entirely underground and brought water into the city from a group of springs located immediately east, along the Via Praenestina.

The Anio Vetus was the first aqueduct to bring water from the Aniene River in the Apennines, which is fed by karstic springs. This ambitious project took three years (272–269 B.C.) and followed the Aniene River valley as far as the town of Tivoli. From Tivoli, the aqueduct turned southward from the base of the Apennines, went west across the volcanic plateau of the Alban Hills, and finally angled northwest down the volcanic plateau into Rome; there it entered the city beneath Porta Maggiore—a total distance of about 60 kilometers (37.2 miles).

Waters flowing into the city via the Aqua Appia and Anio Vetus satisfied the needs of Rome's population for about ninety years. When supplies became inadequate to support the city's public fountains, private users were removed from the system by "cutting off their pipes." Marcus Aemilius Lepidus and Marcus Fulvius Nobilior, censors from 179 to 174 B.C., let contracts to construct a new water supply, but Livy tells us that the project was blocked by Marcus Licinius Crassus, who would

These are ruins of the Aqua Claudia, which brought water into Rome from karstic springs in the Apennines, near the present-day town of Subiaco. The route covered 72.7 kilometers (45 miles) and needed to be precisely surveyed because water flow was driven only by gravity. You can see remains of aqueducts in the "aqueduct park," near the eastern suburb of Cinecittà. The aqueduct came into the city at Porta Maggiore, where there are still standing remnants. Modern aqueducts parallel the routes of the ancient structures and bring water from the same sources.

not allow the aqueduct to cross his property. No new water was brought into the city for another thirty years, until in 144 B.C. the curator Quintus Marcius Rex was charged with restoring the existing aqueducts and building a new one that was to bear his name. The Aqua Marcia brought yet more water from the karstic springs in the Apennines, following very closely the route of the Anio Vetus and also reaching the city at the Porta Maggiore. The Aqua Marcia was a key Roman aqueduct for 680 years until it was destroyed by the Goths.

Between 125 and 19 B.C., the city constructed shorter aqueducts that originated at springs in the Alban Hills volcanic field southeast and east

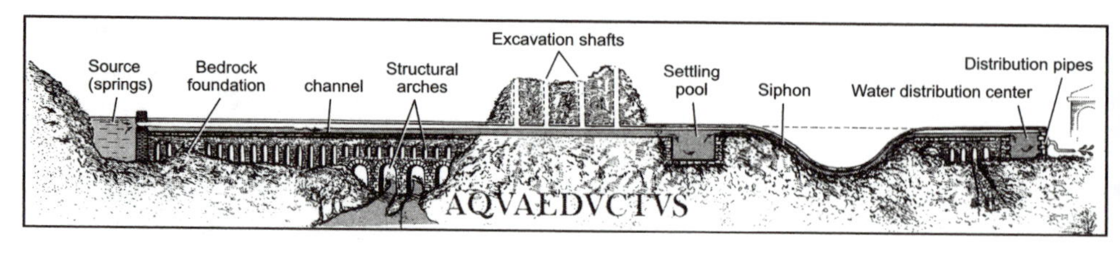

This schematic shows the basic components of a Roman aqueduct, from the spring source to delivery within Rome. (Adapted from DeRosa, Liberati, and Pace 1989)

of Rome. The Aqua Tepula, Aqua Julia, and Aqua Virgo were intended to bring water into a growing city and to fulfill a great demand for water in the public baths. In 40 B.C., Agrippa also repaired the Marcia, Appia, and Anio Vetus aqueducts; to maintain and protect the aqueducts required constant vigilance. In 19 B.C., Agrippa built the Aqua Virgo, which tapped good springs located immediately east of Rome, to supply his thermal baths. After this period, the development of more public baths required either adding new aqueducts or increasing the capacity of existing structures.

After Agrippa's death in 12 B.C., all aqueducts were taken over by the Imperial state. Under Imperial control the aqueducts were maintained initially by 240 public slaves, which Claudius later augmented by 460 Imperial slaves. Maintenance standards were set by the state and enforced by a newly appointed water commission, who were supported by a law enacted in 9 B.C. to protect the aqueducts from vandalism, routine failure, and natural hazards such as earthquakes and floods.

Maintenance was vitally important because the water channels shrank with time as calcium carbonate coated the channel walls. This problem is familiar to modern Roman housekeepers, who attack lime deposits in their sinks with mild acids and vigorous scrubbing. However, instead of a thin scum, the engineers faced deposits many centimeters (inches to feet) thick. Cleaning the aqueduct channels required stripping accumulations from channel walls.

Spring sources located east of the city were tapped by the emperor Augustus, who had the Aqua Alsietina constructed in 2 B.C. to bring water from Lake Martignano, a crater lake located north of Rome to the right side of the Tiber, but the flow was not great, and water quality was poor.

Caligula, in A.D. 38, began construction of two more aqueducts that would deliver water from the bounteous karstic springs of the Apennines. The project was completed in A.D. 52 by the emperor Claudius, for whom the aqueduct was named. The 72.7-kilometer-long (45-mile) Aqua Claudia followed the original route set by the Anio Vetus but often passed through long tunnels rather than canals along hillside contours. The Anio Novus, begun at the same time, also drew water from the river near what is now the medieval town of Subiaco and is more or less parallel to the Claudian aqueducts. This aqueduct carried more water than any of the other classical Roman aqueducts—2,274 liters per second (36,000 gallons per minute)—for a distance of 84.5 kilometers (52.5 miles).

The Aqua Traiana aqueduct, described earlier, was one of the latest of classical Roman times; it brought water from springs located near Lake Bracciano to the Roman neighborhoods along the Tiber's right bank in the district now called Trastevere. Construction of this aqueduct was completed in A.D. 109–10, and you can still travel part of its length either by driving along the traffic-clogged Via Aurelia Antica or by strolling peacefully along the northern edge of Villa Doria Pamphili park.

The Aqua Alexandrina was the last to be constructed by Imperial Rome, during the reign of Alexander Severus (A.D. 222–35). Tapping sources in the tuff plateau east of Rome, the Aqua Alexandrina was 22 kilometers (14 miles) in length.

You'll find many publications that describe the Roman aqueducts—so many that even scholars are not sure how many aqueducts were actually constructed during this period. The estimated number ranges from eleven to nineteen, but authority Thomas Ashby believes that this confusion is based on authors sometimes counting branches of aqueducts as separate structures.

Thomas Ashby's Passion

Three weeks before his death on May 15, 1931, the British archeologist Thomas Ashby left a 925-page manuscript at the offices of Clarendon Press in Oxford. This monumental manuscript, *The Aqueducts of Ancient Rome*, was published in 1935 and is still the standard work on the subject. Ashby's reference book on the engineering and history of the Roman aqueducts was the product of work done in Rome between 1908 and 1925. It is difficult to adequately describe the quality of this elegant study; such a compendium would be difficult to produce today, with budget restrictions, deadlines, publication costs, and a culture of "sound bites." If you have the time and access to a good library, this remarkable work is certainly worth browsing (or even reading).

Aqueduct Restorations

Following nearly 1,000 years of neglect and destruction by invasions and earthquakes after the Empire disintegrated, water supplies to Rome were gradually restored, beginning with the partial reconstruction of the Aqua Virgo, Aqua Traiana, and Aqua Marcia. The Aqua Virgo escaped total destruction because much of it was underground; it was rebuilt by Nicolas V in 1453 and improved by Paul II in 1466–67 to bring water from the slopes of the Alban Hills. The remodeled Aqua Virgo crossed Pincian Hill, feeding several fountains, including the Trevi. Use of water transmitted by the Virgo was expanded to include the fountains and gardens of the Villa Borghese.

Under Sixtus V (Felice Peretti, 1585–90), an aqueduct was constructed to bring water from the territory of Colonna, east of Rome. The Felice was a partial restoration of the Aqua Alexandrina; its sources were in the northeastern slopes of the Alban Hills in volcanic rock. This aqueduct supplied villas on the Esquiline Hill, the area where the Termini railroad station is now located,

During the papacy of Paul V Borghese (1605–21), further work improved the capability of the Aqua Traiana to support not only citizens along the right bank of the Tiber but also "urban agriculture"—the gardens of villas of Monte Giordano and Campo de' Fiori, the Vatican, the Janiculum Hill, and the Villa Pamphili. This expansion was possible because the Aqua Traiana–Acqua Paola brought water from Lake Bracciano.

In 1870, in the time of Pope Pius IX, the new Pia-Marcia aqueduct once again used the plentiful karstic springs that were captured in

Roman times. This recent water system collected 5,000 liters per second (79,200 gallons per minute), which was impressive, but far less than the 8,850 liters per second brought to the city by Imperial Roman engineers.

Water in Modern Rome

That Rome is the only large city on Earth supported in a sustainable way by groundwater can be credited to its unique (and fortunate) coincidence of rock types, regional deformation during plate movements, alteration of limestone by solution, rainfall and snowfall in the Apennines, and Roman engineering. Water sources today are almost the same as those developed during the Empire, and the routes followed are similar to those of the ancient aqueducts.

The most modern aqueduct reaching Rome is the Peschiera-Capore, which brings water from karstic springs in the Apennines over a distance greater than any of the ancient Roman networks. Springs at Peschiera, just east of Riete along the Velino River, are located at an altitude of 410 meters (1,345 feet) and provide a uniform output of 18,000 liters per second (285,100 gallons per minute). In addition, water is drawn from the Capore springs in the Sabine Mountains at a rate of 5,000 liters per second (79,200 gallons per minute). The Peschiera-Capore system was built during successive phases, beginning in 1935, in response to a growing Roman population. This system enters the city from the northeast, via aqueducts that cross the Sabatini volcanic field, to enter the western part of the city, as well as from the Apennines into the eastern part of city along the left bank of the Tiber.

The essential distribution system for water once it reaches the city hasn't changed much since Roman times; it is tied closely to the legacy of the ancient Roman system of spring sources and aqueducts. The simple but sophisticated process involves a tank for maintaining even flow and pressure to a storage tank and then on to a water distribution network. Nearly all the ancient construction was of stone, brick, concrete, and pozzolan cement. The modern equivalent looks different, is made of steel, and involves a type of water tower, called the *piezometer*, that provides an even flow and pressure to storage reservoirs and distribution networks.

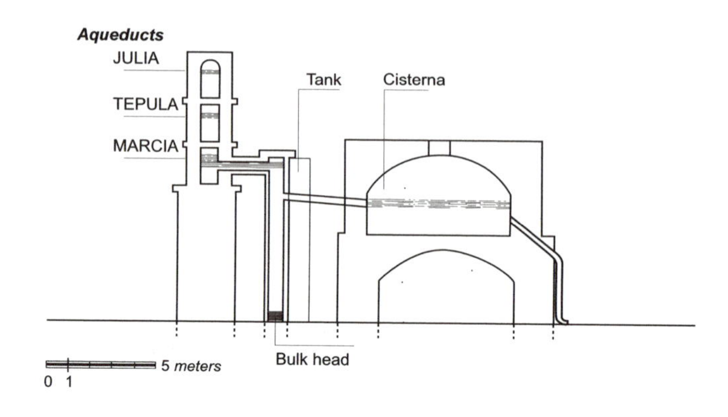

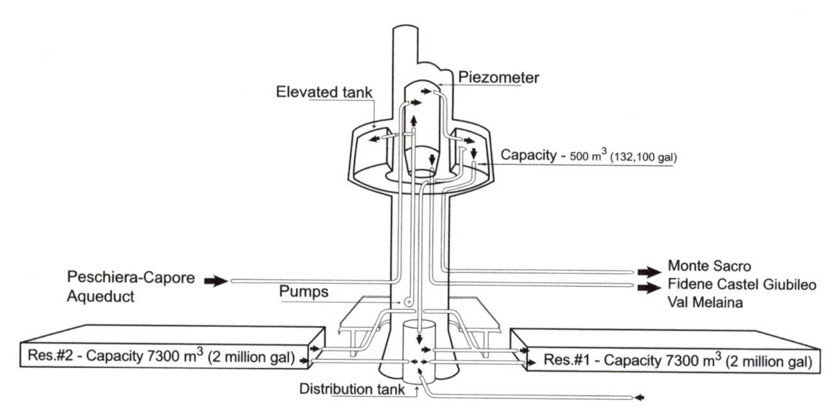

Basic distribution systems don't change drastically. *Top*: Reconstruction of an Imperial Roman cistern located at a suburban villa (by Sartorio 1986). *Bottom*: In a modern "piezometer" in Rome, where the water pressure is controlled, the water is pumped into the elevated tank from an aqueduct, then distributed to the city and to two storage reservoirs.

Materials have changed, but there is a continued sense of history. For example, in 1938, the water authority for Rome constructed a reservoir on the Janiculum Hill, along the Passeggiata del Gianicolo, to hold drinking water from the Peschiera-Capore system. The location is in the Gianicolo Park above the Tiber, from which there are superb views of Rome. The reservoir is not visible, camouflaged as it is by stone walls, shrubs, and a building facade that is a copy of Michelangelo's Roman house. The reconstruction of this famous home fooled one of the authors for two months, until he looked into the window as he was passing by and found not Michelangelo's living room but a gallery with pipes and pumps!

Bottled Water

Rome has great tap water. Why, then, do Romans order bottled water with their meals? Many of the bottled waters are believed to have therapeutic qualities, and many have a unique taste—thus, ordering your favorite water is much like ordering your favorite wine. An Italian law, enacted in 1919, requires that bottlers of mineral water must use a scientific evaluation of the therapeutic and hygienic qualities, along with independent analyses of salt content. The bottle label must contain information about the springs and their chemical and bacteriological content, based on an evaluation by the High Council for Health and university laboratories. Labels on bottled water make interesting reading if you have nothing else to do while waiting for your first course at a local restaurant. Bottled waters, evaluated by total mineral content, are ranked as follows:

Oligominerali	less than 200 milligrams per liter; except for taste, this is drinking-water quality
Mediominerali	between 200 milligrams and 1 gram per liter
Minerali	over 1 gram per liter, which should be used for therapeutic purposes

Bottling water is a thriving business. For example, the Appia brand, from springs in the Alban Hills near the Ciampino Airport, bottles nearly a million units a day. The Claudia brand is from springs near Lake Bracciano, where bottling began in 1897; this company produces nearly 250,000 bottles a day.

The Future of Rome's Water Supply

Most of the springs within Rome have long been buried under tens of meters of rubble that make up the layers of history. Over the last several centuries, factories and businesses have also polluted local aquifers with organic and nonbiodegradable chemicals. Fortunately, some of Rome's springs have been preserved because of their location, such as in the botanical gardens on the slopes of Monteverde. For these reasons, the

This label from a bottle of Ferrarelle mineral water illustrates the remarkable amount of information that you may consume along with the water, including temperature at the spring, the pH, conductivity, and the chemical constitutents. It also shows where and when the analysis was made. Fascinating reading.

springs within Rome are not a major water source for the city, but they remain important as symbols of the sustainability of the city's culture and environment.

Of greater practical concern are those water supplies that originate in the volcanic fields and the central Apennine Mountains. These resources, especially the karstic aquifers in the Apennines, are extremely large and should be able to supply Rome for the foreseeable future; however, there has been little planning to protect the aquifers from pollution in the recharge areas. Until recently, these areas were wilderness or hosted small farms, and there was little to worry either the environmentalists or Rome's water utility boards. Now, following rapid growth of the small villages as sites for summer homes, commuter communities, or ski areas, there is serious potential for pollution of the aquifers, especially by sewage. If this happens, Rome's unique and wonderful water supply will vanish.

• • •

After wandering across the Celian Hill south of the Colosseum, you can return to the arena and from there continue to explore the links between geology and the history of Rome by visiting the largest of the seven hills: the Esquiline.

CHAPTER **9**

Largest of the Seven Hills

THE ESQUILINE (ESQUILINO)

WITH AN AREA of nearly 70 acres, the Esquiline Hill is the largest of the seven hills and has been prime real estate throughout history. You see the edge of this flat-topped hill as a steep slope that rises above the Colosseum and continues to the Basilica of Santa Maria Maggiore.

The Esquiline Hill was the site of Maecenas's gardens, laid out early during the reign of Augustus by Gaius Maecenas, an Etruscan nobleman who left the property to Augustus in 8 B.C. It is said that Nero observed the great fire of A.D. 64 from a tower in the gardens. He must have liked the viewpoint, because after the fire he took over much of the hill, the valley where the Colosseum is now located, and parts of the Palatine Hill for his Golden House (Domus Aurea). Following Nero's death, most of the Golden House was razed when the site was developed as the Baths of Trajan; however, underground portions of the Esquiline wing of the Golden House were rather easily excavated in the well-consolidated tuffs that make up the seven hills.

The Parco di Traiano (Park of Trajan) occupies much of the Esquiline Hill's summit, which you can reach via any of many stairs and streets northeast of the Colosseum and the Colosseo Metro station. The present-day Viale del Monte Oppio cuts diagonally across the middle of the baths, which were enormous when constructed in A.D. 104–9 AD by Trajan's architect, Apollodorus of Damascus. The main bathhouse block alone occupied six and a half acres. In the Parco di Traiano, you can still find some remains of the baths, most notably the semicircular exhedrae, several walls, a fountain house, and a cistern (the Sette Sale).

Public baths like these required a lot of energy: vast quantities of water were heated, food prepared, public rooms made comfortable, and lights provided at night. Trajan's architect began by using essentially free energy from the sun. The hot rooms of the baths were oriented toward the southwest for maximum solar heating, doubly important because historical records tell us that most Romans bathed in the afternoon. The porticoes of many private Roman homes were also oriented to catch the winter sun yet allow shade during the summer months. These simple concepts, so commonly understood in early times, should and could be employed more effectively in our 21st-century homes, as our energy demands exceed energy resources.

Energy Consumption in Rome

All cities require a constantly increasing supply of energy to survive. The amount of energy used in an urban environment is a function of the complexity of people's lives, the degree to which the city's infrastructure is centralized, and the luxuries and conveniences offered to its residents. Citizens of Imperial Rome generally lived near their place of employment, whereas today few people can avoid the dreaded commute through gridlocked streets. Many activities in ancient Roman times were communal, such as bathing, baking, and food preparation; today everyone has domestic hot water heaters for personal cleanliness and kitchen appliances such as stoves and refrigerators for food preparation and preservation. Transportation 2,000 years ago was by foot and boat or, if you were wealthy, chariot or carriage. Today we move rapidly (at least some of the time) by bus, automobile, train, and airplane. While wonderfully convenient and civilized, all these changes come at a tremendous cost to our pocketbooks and to the environment.

Finding reliable information on energy consumption today is not easy, but to do so for ancient Rome is truly a challenge. Many clues to energy resources of the past come from J. G. Landels's *Engineering in the Ancient World*, which is recommended reading for anyone interested in ancient Roman technology—factual, well-written, and humorous, it is fun to read.

The most important difference between energy use in ancient Rome and modern Rome (as an example of urban areas of our time) is that in the past, there was essentially no reliance on geologic resources, whereas

TABLE 9.1
Energy for Rome: The Contrast between Ancient and Modern

	Imperial Rome	*Modern Rome*
Energy Uses	Heating water Transportation Lighting Cooking Smelting, blacksmithing Kilns (ceramic, charcoal) Glassmaking Heating	Heating water Transportation Lighting Cooking Industry Environment control (heating and cooling)
Energy Sources	Wood Charcoal Olive oil (for lamps) Tallow (for candles) Passive solar (in some large public facilities) Coal (only in Roman Britain Labor (humans and draft animals)	Petroleum products (gasoline, diesel, kerosene, jet fuel) Natural gas Coal Electricity (thermal and hydropower) Wood (pizza ovens) Solar (rare) Wind (rare) Geothermal
Domestic / Imported Energy	Wood and charcoal (local, regional, and imported) Olive oil (mostly imported) Labor (local and imported)	Petroleum products (mostly imported) Natural gas (Italian and imported) Electricity (90% imported) Wood (local and regional)
Per Capita, Annual Energy Consumption	Precise amount not known, but subjective logical observation indicates it was very little	[Includes industrial use] Electricity—28,000 kilowatt hours Petrol—292 kilograms Diesel—690 kilograms Heating oil—20 kilograms Natural gas—371.6 cubic meters Wood for pizza ovens—??

Source: Data from Provincia di Roma, 1997.

today we rely almost entirely on fuels extracted from the Earth or from hydropower dams that also depend on the Earth's water supplies. Any comparison of energy use in ancient and modern Rome must be subjective, but this table provides a rough contrast. It also is a reality check and an indication of how imperative it is that we both change our lifestyles and redesign our cities: we must find ways to achieve a sustainable existence, while reducing our energy dependence and without plunging ourselves back into the very basic existence of the ancient Romans.

Natural Energy Resources in Ancient Rome

Apart from the olive oil burned in lamps, two fuels were used in Rome: wood and charcoal. Charcoal was generally preferred for cooking because it burns slowly and generates less smoke (air pollution was a problem in ancient Rome as it is today). Charcoal burns at a temperature of 900°C (1,652°F), but it can reach 1,300°C (2,372°F) when a bellows forces air through the burn. The higher temperatures also made charcoal the fuel of choice for metal-smelting furnaces.

Charcoal is formed when carbonaceous materials such as wood are heated and partly burned. The heat drives off water and volatile substances and breaks down the carbohydrates. Closely packed piles of wood are burned slowly in kilns to produce charcoal, which is easier to ship because it weighs only 25 percent as much as the original woody materials.

Forests were once lush around Rome. At lower elevations, there were evergreen oaks and coastal pines. In the foothills of the Apennines and on Rome's volcanoes were abundant deciduous trees, including oak, hazel, maple, elm, sycamore, cypress, pine, and some juniper. The cypress so commonly seen here now was introduced from Greece. The best charcoal for Rome was generated from ilex (holm oak) and beech trees. J. Healy, in his book on Greek and Roman mining and metallurgy, calculated that 5,240 hectares (13,392 acres) were deforested each year within the Roman Empire to produce charcoal for smelting iron. The Romans also imported charcoal from Magna Graecia (the Greek colonies of southern Italy), Macedonia, Mount Ida (Crete), and Gaul (modern France).

As the population of Rome increased, so, too, did the need for fuel wood and charcoal for heating (charcoal braziers), cooking, industry (metal smelting), construction (fired tile and brick), and cremation. Funeral pyres alone consumed great quantities of wood—preferably oak or other hardwoods. At the end of the 1st century A.D., burial of the dead began to replace cremation, not as a consequence of wood shortages but due to cultural changes. This practice reduced the demand for firewood temporarily, but demand increased once again with the larger market for fired bricks and fuel needed for heating the thermal baths. By far the greatest demand for fuel wood resulted from the popularity of *thermae* or thermal baths, beginning in the 2nd century B.C. and reaching a peak in the 4th century A.D. We read that in A.D. 364, for African ship owners to retain special privileges in Rome, they had to carry logs from North Africa in addition to whatever cargo they were delivering. Russell Meiggs, in his book on Mediterranean forests, assumes that such tributes were required of the merchants because forests around Rome had been exhausted.

Using their reconstruction of the plumbing system of the Baths of Caracalla, Leonardo Lombardi and Angelo Corazza estimated how much energy was required to maintain the baths. Fires below concrete tanks heated bathwater and produced hot air that circulated into the baths, where temperatures were maintained at 50°C (122°F). From the same system, the sauna was kept between 60 and 80°C (140 and 176°F). This system would have required heating a 10-cubic-meter (2,642-gallon) tank of water from the inlet temperatures of 10 or 12°C (50 or 53.6°F) up to the desired 40°C (104°F)—a process that would consume 85 kilograms (187 pounds) of wood for initial heating and 18 kilograms (40 pounds) per hour to maintain. For the Romans, this supply of hot water was important enough to be supported by a tax on the baths and regulation of the wood supply. By law, the public baths had to keep a month's supply of wood on hand. Cleanliness and comfort had their price: citizens often complained about irritating smoke pollution problems generated by the baths.

In private homes during the 1st century A.D. the *hypocaust* used terracotta pipes to carry hot air throughout the building from a wood-fired heater in the basement and was one of the first central heating systems. After the collapse of Imperial Rome, until the Renaissance, the population of Rome could only gather around wood fires on cold winter days or sit in the sun for comfort.

Several other natural resources were used by the Romans in lesser ways, such as the solar energy we discussed at the Parco di Traiano. Water mills, which were rare in Rome because the small streams flowed into the Tiber only seasonally, were used only in the Trastevere neighborhood. The steam engine was a concept promoted by Hero of Alexandria (in the 1st century B.C.), but it wasn't put into use until far later because of either technical problems or the lack of a high-quality fuel.

Another obvious and traditional source of energy for construction was the labor of men and draft animals, which by today's standards was inefficient, even with the assistance of mechanical devices such as the block and tackle and the treadmill. Landels notes in his book on Roman engineering that one gallon of gasoline used by an ordinary engine of 1978 does the work of ninety men or nine horses. Most of the animal power in Rome required strong oxen, although horses were the logical choice for rapid communication tasks.

Beyond the kitchen fire, lighting during the Republic and Imperial Rome primarily involved oil-filled saucers with linen wicks, replaced in part by candles during the 1st century A.D. Nothing much changed until the mid–19th century A.D., when gas lighting was installed in Rome and other great cities of Europe, providing bright, clean illumination. At that time, gas was generated from coal and held in *gasometers*, storage tanks that maintained pressure for the distribution systems. Between 1846 and 1878, Pope Pius IX's modernization of the city included the installation of gas streetlights. Pius's gasworks were constructed in the northwest end of the Circus Maximus, allowing Rome a degree of modernization but, unfortunately, ignoring the glories of the past.

Energy in Modern Rome

As you stroll through Trajan's baths across the Esquiline Hill, look at the outward evidence that we citizens of the 21st century use a lot of energy: cars roar by (if they aren't in gridlock), apartments emit the sounds of stereos and televisions, stores offer the latest in appliances, restaurants are brightly lit and use the latest equipment, and most buildings are heated during the winter or cooled on hot summer days. Today, all the energy used by Romans is imported from outside the city. Within the state are seven hydroelectric plants and two thermoelectric plants (fueled by petroleum products), which meet only 10 percent of the local needs; the rest comes from plants or hydroelectric reservoirs outside the state. In 1993, electricity consumption in Rome was about 7.873 gigawatt-hours, of which 1.8 percent was for public lighting, 38 percent for households, and 59.3 percent for other purposes (mostly industrial). Consumption continues to rise as residents purchase additional electrical appliances and install air-conditioning. In 1993, Rome also consumed 818,940 tons (195 million gallons) of petrol; 1,940,490 tons (467 million gallons) of diesel; 159,910 tons (28 million gallons) of gasoline; 50,370 tons (11 million gallons) of heating oil; and 1,040,370,000 cubic meters (36,700,000,000 cubic feet) of natural gas. There are no similar statistics on how much wood is cut to maintain the ubiquitous pizza ovens, and the ovens definitely do generate smoke and pollution, but no conservation effort is likely to render any change—it is a great smell to stimulate your appetite on a cold winter night, and, after all, tradition is tradition!

One way to grasp the reality of modern energy consumption in Rome and other Italian cities is to look at an image of the radiance observed at night by meteorologic satellites. The visible emissions from cities are evident, outlining the Italian Peninsula and forming large spots associated with the cities. On the Tyrrhenian coast, the two largest foci for energy usage (lighting, heating, etc.) are metropolitan Rome and Naples. Just imagine: if we had similar images from 2,000 years ago when Romans were gathered around fires or going about their business with oil lamps, we would have seen nothing except the thermal signatures of forest fires.

Except for some wood, all modern energy requirements are met with geologic resources, most of them imported. Italy does have some oil and gas fields. The largest is in the Po River basin, with an estimated 0.4 billion barrels of oil and 0.54 trillion cubic meters (18.9 trillion cubic feet) of natural gas. Other areas with some promise include the Adriatic basin and the Apulia platform, but even if they were brought on-line, most of Italy's petroleum would have to be imported.

In this nighttime image of the Italian Peninsula, assembled in the 1990s from U.S. meteorologic satellites, the bright spots are visible emissions from cities. Urbanization of the Italian coastline is evident; the broad patches that spread inland are associated with metropolitan Rome and Naples. These images correlate well with energy use and, indirectly, with population. Most of this energy is directly or indirectly derived from petroleum products. (From The U.S. Air Force and National Oceanic and Atmospheric Administration's National Geophysical Data Center)

Italy and Rome are not alone in their usage scenarios: when we return home, we can judge for ourselves how our own nations, cities, and communities consume resources by examining satellite images of our own areas and from data provided by local power companies. In the meantime, let's return to Rome's streets.

• • •

If the weather is warm and clear, continue north along the surface of the tuff plateau of the Esquiline Hill toward Santa Maria Maggiore and the Termini Station, where further historical and cultural treasures await you. If it is cold or raining, you have a wonderful excuse to establish yourself in a coffee bar and contemplate Roman life. At this point it may be hard to get excited about the geology; the plateau surface here would be reasonably featureless anyway, but since it is covered with the debris of time and buildings, it is very difficult to envision. Perhaps you must take our word for it—this area is fascinating when you integrate all that we know from ancient excavations and underground quarries, modern excavations for foundations, and the results of test drilling.

CHAPTER 10

Upper Class

THE VIMINAL (VIMINALE) AND QUIRINAL (QUIRINALE) HILLS

It is sometimes difficult today to visualize each of the seven hills of Rome, and if you become confused, you aren't alone. In fact, the confusion goes back several millennia: from the 1st century B.C. to the 1st century A.D., the number of "hills" identified in historic documents vacillated between six and eight.

You may remember that earlier we described the flat-topped summits of the Quirinal and Viminal hills as the surfaces of pyroclastic flow deposits (tuffs) from the Alban Hills volcanoes. At the base of the hills, the tuffs of the tableland overlie deposits of sands and gravels left by the ancient Tiber River before they were buried by volcanic ash.

The Viminal Hill

The Viminal Hill is a mere sliver of a plateau left between the early creek beds that are now occupied by the Via Cavour and the Via Nazionale. The creeks originated in poorly defined headwaters near today's Piazza della Repubblica. Just try to imagine bubbling brooks flowing down two of the busiest streets in Rome!

A rather subdued plateau, the Viminal Hill is difficult to identify because of its closely spaced buildings and narrow streets; however, in the broad, open area of the Piazza dei Cinquecento, near the train station, you have a general idea of the appearance of the original plateau surface. The head of the Viminal plateau was the site of the Baths of

Diocletian, which occupied more than 2.5 acres between the present Piazza dei Cinquecento and the Piazza della Repubblica.

The most easily recognized summit on the Viminal Hill is the foundation beneath the Ministero dell'Interno, which is best seen from where you enter the Piazza Viminale from the Via A. DePretis. Another familiar public building on this hill is Rome's Opera House. If the opera ever produces Gounod's *Philémon et Baucis*, Jupiter, sung by a baritone, would be right at home here—the hill may have been named after Viminus, a deity who was a form of Jupiter.

As you walk northwest from the Viminal to the Quirinal Hill, you are dropping into a stream valley and climbing out again. This ancient creek bed underlies the Via Nazionale, then curves southward along the Via dei Serpenti as far as the Via Cavour.

The Quirinal Hill

In Imperial Roman times the Quirinal was an upper-class residential area, and even today the hill is dominated by the Palazzo del Quirinale, the official residence of the president of Italy. Until 130 years ago, this was the pope's summer palace, and then for 76 years it was the palace of the Italian royal family. The Quirinal is full of treasures and surprises. At the summit is the palazzo, which dominates central Rome as it rises above the crowded quarter that includes the Trevi Fountain. Your visit to this area will be rewarded with a rare sight—a piazza without parked cars! The piazza has a good view across the city to the east, but your eyes will be drawn by the enormous sculptures of Castor and Pollux, brought to the hill in 1588, as well as a granite obelisk, which was installed in 1786. Although the important Acque Sallustiane springs, once active in the valley between the Pincian and Quirinal hills, have been buried by the evolving city, one of them actually flowed for a time in the basement of the Hotel Bristol Bernini, near the Piazza Barberini.

If you follow the Via del XXIV Maggio to the Largo Magnanapoli and turn southwest toward the Roman fora, you'll have an excellent view of Trajan's Column as you descend the steps between two medieval buildings.

• • •

In the valley of the Roman fora are Trajan's Column, Forum, and Markets. Adjacent to the Markets of Trajan are the Forum of Augustus, with its Temple of Mars Ultor (2 B.C.), and the Forum of Nerva (A.D. 97), both excellent examples of the way Romans used locally quarried tuff blocks for construction (perhaps from the Quirinal Hill). These well-consolidated but easy-to-quarry blocks have resisted the elements and earthquakes for several millennia. They did not, however, resist looters, who have removed metal, marble, and travertine facings. You can see holes in the tuff blocks that may have been drilled for tangs to hold the facings.

The foundation for Trajan's Column was excavated in sandstone deposits of the ancient Tiber, near a contact with the overlying tuffs that make up most of the Quirinal Hill. You'll see why this geologic setting is important as we compare this well-preserved marble column with the similar Column of Marcus Aurelius, which is located only 700 meters (2,300 feet) away but on the Tiber's floodplain. The magnificent columns of Trajan and Marcus Aurelius are almost the same size, are made of the same materials, and were constructed in a similar fashion. The weight of each column is about 1,000 tonnes (1,100 tons), but the shapes are slightly different.

Trajan's Column was erected in A.D. 113 to celebrate Trajan's victories against the Dacians (from modern Romania). A bronze statue of the emperor on its summit was replaced at the end of the 16th century by one of Saint Peter. Wrapping the column shaft is a figurative frieze describing Trajan's military campaign; the carvings were executed on seventeen cylindrical marble blocks from Luni, each with a diameter of 3.6 meters (11.8 feet) and a height of 1.5 meters (4.9 feet). The column base is 12 meters (39 feet) high and rests directly on lower Pleistocene sandstone.

The bas-relief on the Column of Marcus Aurelius tells the story of the German and Sarmatian wars of the "emperor-philosopher." At the end of the 16th century, his statue, too, was replaced with one of a saint—Saint Paul. Near the middle of the column, the marble cylinders have been dislocated about 10 centimeters (4 inches), decapitating several Roman soldiers and a winged Victory. There has been a vigorous

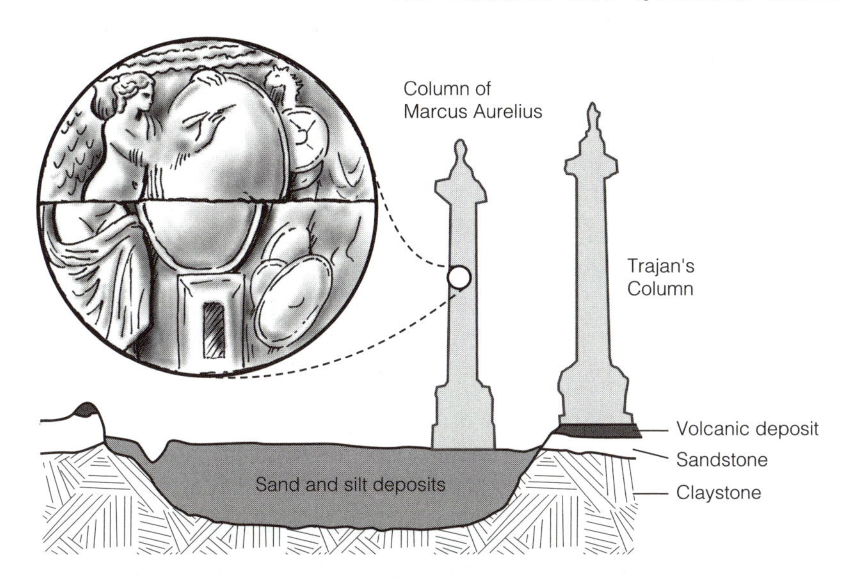

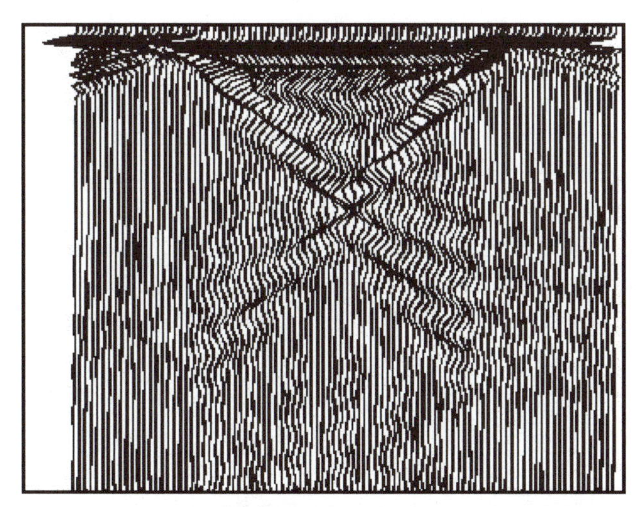

To compare the damage to two triumphal columns of similar size, age, and construction, look at Trajan's Column and the Column of Marcus Aurelius. The damaged Column of Marcus Aurelius was sited on poorly consolidated river sediments of the Tiber, and the intact Trajan's Column was built on marine sedimentary rocks and volcanic tuffs. A model of ground-motion amplification in the sediments of the Tiber illustrates substantial soil movement below the Column of Marcus Aurelius, where seismic energy was "trapped" in the ancient river channel, and the illustration shows an offset in the marble cylinders that make up the column.

debate about the origin of this dislocation. Experts have determined that the column was damaged by one or more earthquakes, after thieves had removed metal clamps that were needed to stabilize the marble cylinders, sometime after the 13th century and before a structural refitting of the column ordered by Pope Sixtus V in 1585. The only strong earthquake that had affected Rome during that period occurred in 1349 and was most likely the culprit.

One aspect that supports this hypothesis is the difference between geologic foundations under the columns. Trajan's Column is situated on claystone and sandstone, and the Column of Marcus Aurelius rests on 60 meters (197 feet) of unconsolidated alluvium left by the Tiber River during the last 12,000 years. Antonio Rovelli, of the National Geophysical Institute, simulated the ground acceleration that would accompany a strong earthquake near both of the columns. Rovelli's simulations indicated that the sandstones below Trajan's Column were more rigid and therefore not vulnerable to increased ground acceleration that shook the river sediments underlying the Column of Marcus Aurelius.

Another factor that would have increased the earthquake damage is the shape of the sediment-filled ancient valley of the Tiber. To confirm Rovelli's calculations of ground response, the Institute of Seismic Engineering in Skopje, Macedonia, constructed a 1:6 model of the Column of Marcus Aurelius and placed it on a vibrating table. Moderate horizontal ground acceleration at the base of the model damaged it very much like the actual Aurelius column. Scientific conclusions based on both the numerical modeling and engineering tests indicate that the Column of Marcus Aurelius was damaged by a strong earthquake with an epicenter about 100 kilometers (60 miles) from Rome and that amplified ground acceleration was a consequence of the column's location on the unconsolidated sediments. This practical research and the evidence we see in comparing the two columns are ample proof that understanding the underlying geology is vital when planning the construction of any structure.

While you are on the Quirinal summit, spend some time in the "pearl of the baroque," the Sant'Andrea al Quirinale—one of Bernini's most exquisite creations, where you can appreciate the artist's use of the various marbles found in Italy. In almost every part of Rome you will see the influence of Gian Lorenzo Bernini, the baroque artist whose prolific output during seventy active years encompassed all media, including painting, fireworks displays, and stage design. He is best known for his mastery of marble, turning stone into living forms. His marble sculptures range in type from the remarkably detailed busts of Cardinal Scipione Borghese, a patron, to the larger-than-life-size *Apollo and Daphne*, where the leaves growing from Daphne's fingertips are so delicate that they are translucent. How could one stone have such strength yet allow the carving of fine detail?

Marble: The Stone Carver's Palette

Too closely examining or touching any of the Bernini sculptures in Rome will certainly earn a strong rebuke from a museum guard, but fortunately, it's possible to explore the textures of marble columns in Sant'Andrea al Quirinale and other churches. Look closely, and you can see that marble is crystalline. The calcium carbonate (calcite) crystals that make up marble range in size from those that are nearly invisible without the aid of a magnifying glass to those the size of a fingertip. If you compare marble with limestone from building blocks used in monuments and buildings you've visited, you'll discover that although the chemical compositions are the same, their textures are very different. Limestone, although lithified (turned into rock), shows the original depositional conditions of limy sediments, ranging in depositional environment from sea floor muds to the fossils of a coral reef. Metamorphosis of limestone to marble usually erases the original rock texture, destroying any clues to the original deposition environment.

When we say marble is a metamorphic rock, we refer to the process of change, during which the original rock, in this case a limestone, recrystallizes in response to increased pressure and temperature as it is progressively buried in sedimentary basins by younger deposits. Some of these basins are as much as 4 to 5 kilometers (2.5 to 3 miles) deep, and the weight of overlying rock creates extreme pressures. These rocks can be taken even deeper if they are located along a descending crustal plate; at a depth of 10 kilometers (6.2 miles), the pressure is equivalent to 3,000 bars (2,961 atmospheres at sea level) or 42,935 pounds per square inch. The rock doesn't melt, but much of the original texture is destroyed as interlocking calcite crystals grow. Uniformity of crystal size occurs where there is the right combination of original rock type, pressure, and temperature; the more uniform the crystals, the better the marble is for carving.

Although it is a common perception, not all marbles are uniform and white. There is a wide variety of colors and textures, many of which you can see in the polychromatic marble inlays of floors from all of Rome's historic periods. The colors may be related to the composition of the original sediment, in which mineral impurities or organic materials have stained the rock. For example, just a small amount of ferric iron oxide can provide the warm red colors often sought for marble facings and columns. Organic materials in the original limestone can produce black marble—you'll see an excellent example in the stone used for Bernini's *Tomb of Sister Maria Raggi* in the Church of Santa Maria Sopra Minerva.

Texturally, many marbles are far from uniform. In some of the marble columns seen throughout Rome, the rock looks as if it had been chopped up and reassembled. In some respects, that is exactly what happened, except that the processes were natural. Marbles composed of many angular fragments are called *marble breccias*. Breccias are rocks made up of angular pieces of more than one rock type; they have several origins, the most common occurs when brittle rock is shattered by faults. Water flows through the open space between fragments, dissolving some of the marble, and eventually the open spaces are filled with precipitated calcite. The final product is a rock that appears fragmented but can be quarried and carved. The multiple colors come from the variety of marbles that were broken or from staining of the calcite that cements the fragments together. Many modern faux marble floors in Rome are similar in appearance to breccias but are man-made—by mixing a variety of marble chips with cement and then cutting the product into slabs.

Another marble texture is created when marble is sheared while under pressure. This type doesn't have angular fragments; instead, it looks more like fluid bands or layers that pinch and swell across the carved column. In these cases, the rock has been deformed plastically, not brittlely, by a combination of shearing and solution. This process can occur either at great depths (more than 10 kilometers) or close to active faults, where bands of ductile rock under pressure were squeezed like toothpaste from a tube. Some marble textures include regular or very irregular white veins—calcite that was precipitated along fractures opened by shearing rock or escaping gases. Veining can occur at many stages during the rock's metamorphosis.

Vatican Treasures Provide Clues about Geologic Processes

By carefully examining marble textures and having some knowledge of where the quarry is located, it is possible to put together the rock's tortured history. One of the authors (Funiciello) was doing just that when he took a class studying structural geology to Saint Peter's Basilica. The university class was studying the geologic structures evident in columns of Cottanello marble. The students' unusual behavior, describing and making measurements on marble columns in one of the world's best-known basilicas, drew the immediate attention of Vatican Security. Not knowing what to do with this group, the guards asked Funiciello to come with them to the office of the Swiss Guard, then consulted the man in charge of antiquities for the Vatican. Don Vittorio Lanzani, secretary of Saint Peter's Basilica, was delighted to learn about the research, which would provide more insight into the origins of the Vatican marbles, and an open invitation was given to study the columns anytime (with some advance notice, however!).

Nearly every imaginable marble texture and color can be found in Rome. Beginning in the 1st century B.C., marble was imported into Rome on a large scale, for veneers, columns, and statuary. Much of the marble came from the Roman colony of Luna (modern Luni), near Carrara. Marble quarries became state property, and soon column shafts were mass-produced in standardized heights of 24, 30, 40, and 50 Roman feet (7, 9, 12, and 15 meters; or 23, 29, 39, and 49 in modern foot measurements). The stone came by raft up the Tiber from the harbors near Ostia to the Emporium, also called Marmorata, a district of warehouses and shops located south of the Aventine Hill; part of the trip took place along what was then a tributary of the Tiber that underlies today's aptly named Via Marmorata: *marmo* is the Italian word for marble. Marble was stockpiled in Marmorata and sometimes not used for centuries. As late as the 20th century, a few pieces from the stockpile were sold to J. Paul Getty for his reconstructed Roman villa in Malibu, California. The early use of multicolored marble breccia from Africa, white-and-black "antique" marble breccia from the French Pyrenees, and white-and-red marble breccia from Asia Minor "pavonazzetto," among others, reflects not only the Romans' eye for the aesthetic properties of varied stones but also their ability to quarry and transport the material to Rome from around the Mediterranean.

How do we keep track of the bewildering array of marbles? Ciriaco Giampaolo and his colleagues, of the University of Romea Tre, have established a database for stone used throughout Italy. Different marbles exhibit distinctive geologic "fingerprints" that can be identified in the laboratory. The database includes not only what has been a qualitative assessment of the stone source but also a geologic and chemical characterization of each type, comparing it with what is known about the original quarries. The Empire's passion for stone led to the import of material from all corners of the Roman world. Giampaolo notes that in some cases, the demand was so great that deposits of some particular stones were completely depleted, leaving nothing even for subsequent repair of monuments. In later periods, marble came from older Roman buildings that were themselves "quarried."

Carrara Marble

Some of the world's best marbles are part of a crumpled wedge of metamorphic rock shoved to the surface of the Earth as the Alps collapsed into the crust. Sited on high, precipitous peaks in northwestern Italy, the large Carrara quarries have been active for thousands of years and still provide much of the best-quality marble to the 21st-century world. Carrara's quarrymen and stoneworkers, known throughout the world for their talent and experience, use both modern and traditional methods.

Carrara marble has been used in Rome through time because of its ivory white color, uniform grain size, medium hardness, and ease of working with a variety of stone-carving tools. The quality of the marble and the people who worked it were well known to Gian Lorenzo Bernini, who chose it for numerous sculptural pieces. As well as the pure, homogeneous Carrara marble, Bernini employed a broad spectrum of colored, textured marbles, including Cottanello marble. This wealth of material was available from around the Mediterranean and provided Bernini with the marble palette he needed.

The Cottanello Marbles

Rich complexity and *visual warmth* are words that come to mind when describing the sheared and veined red-and-white marble that was used for interior columns in the Basilicas of Sant'Andrea al Quirinale, Saint Peter, and Sant'Agnese in Agone (on the Piazza Navona). This is Cottanello marble, from the Sabine Mountains, and was quarried in ancient times and again by Bernini during the artistic explosion of baroque Rome.

The Cottanello quarry is located along the Sabine fault, a major strike-slip fault that runs north-south from the Apennines toward the Alban Hills. Limestones cut by the fault have been deformed over millions of years. At great depths, the rocks were ductile (stretchable) along the fault, producing sheared and veined material. More competent (stronger) material, mixed with clay and marl, added to the complexity of the final rock, in which the original sedimentary bedding has been completely masked by the metamorphic processes of pressure, shearing, and heat. Two of the Cottanello columns in Saint Peter's Basilica show less shearing and recementation of the rock, so structural geologists can determine that they were quarried at different distances from the fault.

The Cottanello quarry is now abandoned, but in the quarry floor two partly excavated columns remain, each still showing the quarrymen's tool marks as if tomorrow the workers might return to finish their job.

• • •

You can leave the Quirinal Hill via steps descending from the Piazza del Quirinale. Now turn north and then east to where we began our journey—at the Trevi Fountain. Find a bar or a *gelateria*, sit down to enjoy your drink or ice cream, recall Anita Eckberg's famous romp in

The variegated marble columns of the Basilica of Sant'Agnese in Agone are Cottanello marble, quarried in marble deposits adjacent to a significant strike-slip fault that cuts down the spine of the Sabine Mountains northeast of Rome. (Funiciello and Mattei 1991)

the fountain, and watch the appreciative crowd enjoy the splashing waters of this elegant symbol of Rome. You have now journeyed through time and space in this great city, learning about the role of its geologic setting in the creation of its precious artworks, the engineering of its infrastructure, and the development of the remarkable Empire itself.

The historical Cottanello quarry, located close to a large strike-slip fault in the Sabine Hills. In the foreground is a partly cut column left in place by Bernini's stoneworkers sometime during the 17th century. Stone columns were traditionally rough cut in the quarry because transport was easier without carrying along the waste.

CHAPTER **11**

Field Trips in and Around Rome

GEOLOGISTS love to lead field trips for other geologists. It is a great way to expand your knowledge of the field and to make new acquaintances. In most cases, the trips follow a printed guide that allows others to take the same tour at any time. These guides are valuable in that geologists rarely finish a trip on time; vigorous debate at each stop usually extends the trip until the last daylight has faded away. We have prepared three field trips for the reader. The first trip is entirely on foot. The second and third trips require some public transport or a rental car.

Recommendation: The streets of Rome follow the natural terrain and not a rectangular grid. We recommend that you carry a small compass and a detailed street map while visiting the city (or, if you are technically minded, a small global positioning system [GPS] receiver and digital map).

THE SEVEN HILLS OF ROME IN FIFTEEN STOPS

THE SEVEN HILLS of Rome—Quirinal (Quirinale), Viminal (Viminale), Esquiline (Esquilino), Celian (Celio), Capitoline (Campidoglio), Palatine (Palatino), and Aventine (Aventino)—have been important to the history of this complex city. The original geologic framework of Rome is evident when seen from the ridge overlooking the right bank of the Tiber. From the vantage point of the Janiculum (Gianicolo) Hill, the nearest terrain feature beyond the Tiber is the tuff plateau that extends from central Rome eastward to the base of the imposing volcanic edifice of the Alban Hills. The plateau surrounding the Alban Hills was constructed during explosive eruptions that occurred sporadically between 550,000 and 350,000 years ago. The Roman landscape was shaped during the deposition of hundreds of cubic kilometers of volcanic ash and pumice that covered the future site of Rome, altered the course of the principal rivers, wiped out all forms of animal and plant life, and constructed sloping plateaus. These geologic features were the setting for development of Roman culture and civilization. The periphery of the tuff plateau, in proximity to the river, was eroded to form small, isolated hills that were easily adapted for shelter and human settlement.

The seven hills, carved from the sloping tuff plateau by tributaries of the Tiber, are no more than 60 meters above sea level and relatively small. The plateau overlies continental sedimentary rocks of the Pleistocene Age, which are now exposed in the area of Ponte Galeria, west of the city.

Local climatic and hydrologic conditions at the future site of Rome were particularly favorable for rapid incision of the tuff deposits to form valleys along which there are outcrops that allow rare glimpses of the geologic structure underlying the city. The purpose of this itinerary, to be followed on foot in half a day (it is easiest to do this on a holiday or Sunday morning), is to understand the connection between geology and the city's evolution from ancient to recent times.

Stop 1. Janiculum (Gianicolo) Hill

We begin on the Piazzale Giuseppe Garibaldi on a terrace overlooking Rome from the Janiculum Hill. The best time to see the geologic structure and the urban landscape is on a clear day during the early morning or late afternoon. The tuff plateau extending out from the Alban Hills volcanic field is on your right. Directly in front of you, beyond the city, are the rugged Apennines, represented by the Lucretili, Tiburtini, and Prenestini mountains. Look carefully, and you can see the smooth surface of the tuff plateau, incised near the Tiber by ancient streams that left the subtle relief characteristic of the famous "seven hills." The lowest part of the city (from the base of the Janiculum as far as the "seven hills") is on the recent alluvial plain of the Tiber. Now, walk southeast along the Passeggiata del Gianicolo to the Acqua Paola.

Stop 2. The Acqua Paola Fountain

The Acqua Paola Fountain is the ornate terminus of the Roman emperor Trajan's aqueduct (the Aqua Traiana), which was rebuilt by the order of Pope Paul V to bring water from Lake Bracciano. These "oligominerale" waters issued mostly from aquifers in volcanic terrain and were more insipid than the bicarbonate-alkaline waters provided by

Views across Rome, looking east from the Janiculum Hill. In the first two views, the background is dominated by the Apennine Mountains, the source of much of Rome's water. In the middle ground is the dissected tuff plateau of the "seven hills of Rome." Without partial burial by debris that has collected over the millennia and the buildings constructed on top, this plateau would be easy to identify. That is not the case, however, and the seven hills are marked. Below the hill is the Tiber River, marked by lines of trees along its banks.

most of the Roman aqueducts. As a consequence of disparaging remarks by the populace considering the quality of the water, the statement "It is from the Acqua Paola" referred to "taking away your taste."

The steep slope between the Acqua Paola and Trastevere's alluvial plain was, in ancient times, used for a system of mills and public hydraulic works that may have been located near the Via di Porta San Pancrazio and under the small plaza of San Pietro in Montorio (the Roman latrine below San Pietro can be visited with permission of the superintendent). Located immediately east of the Acqua Paola is the Spanish Academy of Rome. In 1917, Pablo Picasso prepared the sketches for the work *Parade* from the residence of the Spanish Academy. This work was produced at a time of transition to the most celebrated phase of this extraordinary artist's life. In *Parade* the skyline of the Alban Hills is visible, along with the Roman landscape of the fora, as it would have been seen from the terrace of the Spanish Academy.

The *fontanone*, or terminal fountain, of the Acqua Paola, built by Pope Paul V, adorns the restoration of the Aqua Traiana to bring water to Rome from Lake Bracciano in the Sabatini volcanic field.

Proceed downhill along the Via di Porta San Pancrazio (steps immediately north of the Acqua Paola) toward Santa Maria in Trastevere and across the narrow lanes that surround the basilica. During your descent you are passing by the Parrasio Woods and the magnificent University Botanical Garden (worth a visit). Where you cross the Via Garibaldi there is a fountain that corresponds to one of the numerous springs that were located at the base of the Janiculum Hill. From here, descend directly by the Vicolo di Frusta, left on the Via della Paglia, then right on the Via della Lungaretta, moving toward the Tiber and reaching Santa Maria in Trastevere at Stop 3.

Stop 3. Santa Maria in Trastevere

The site of Santa Maria in Trastevere occupies the site of an ancient Roman inn associated with a prodigious outflow of mineral water

Parade, painted by Picasso in 1917, while he was in residence in the Spanish Academy, which is across the street from the Acqua Paola. The skyline of the Alban Hills is visible through the window.

with apparent traces of oil, which led to the inscription on the central nave—"Fons olei." (The flow was presumed to contain hydrocarbons but was probably mineralized water rich in sesquioxides and oxides of iron and aluminum, as has been seen in wells for the tobacco factories on the Tiber Island and in the Roman Forum (Acque Lautole). These waters appear to be from a conglomerate layer at the base of recent alluvial deposits. The basilica is on the alluvial plain, and an inscription describing damage to the building may refer to medieval earthquakes of A.D. 1091. Throughout the city, earthquake damage is worse where poorly consolidated sediments underlie buildings. The few records of such an event may have been derived from stories of emergencies that occurred in Rome toward the end of the 11th century.

Proceed east along the Via della Lungaretta to the intersection with the Viale di Trastevere, where it is possible to visit the stately lower basilica of San Crisogono.

The decorated outlet for a spring located at the base of the Janiculum Hill (on the Via Garibaldi). (In common with third field trip itinerary)

Stop 4. "Ponte Rotto"

Continue along the Via della Lungaretta toward the Tiber (Lungaretta follows the old Via Aurelia) to the Ponte Palatino (site of the Pons Aemilius, also called the Ponte Emilio or Ponte Santa Maria). The older bridges were seriously damaged and isolated by the flood of 1598. A

Santa Maria in Trastevere, the first Christian church in Rome open to worship, is located near a mineral spring with a prodigious outflow.

remnant of the last bridge still stands south of the Tiberina Island and is called the "Ponte Rotto." Just after the destruction of the Ponte Emilio during the flood of 1598, the Via della Lungaretta eventually lost its importance as a main thoroughfare. In the 1800s an imposing pontifical edifice was constructed to shelter Santa Maria in Trastevere, and the most fundamentally important road plan in at least 2,000 years was implemented to reduce the narrow rectilinear arteries. The Via della Lungaretta had fulfilled its final role.

Following the 1598 flood, as one would expect, travel by the ancient Via Aurelia was curtailed (the Via Aurelia was one of the principal roads entering Rome from the west). Subsequently, the central area of the Roman fora was heavily urbanized from late medieval to Renaissance times. The urban evolution of Trastevere eventually caused the loss of the Via Trastevere as a main thoroughfare between the west bank and the Roman fora.

The flood of 1598 was part of a flooding cycle that lasted about 300 years and was related to a global climatic event, a time of cooler, wetter weather in Europe. Exceptional river levels, measured at Ripetta, rose to over 14.5 meters (47.6 feet) above the riverbed, spreading silty water through the city many times.

The Via Aurelia once crossed the Tiber at this point. The bridge, now called the "Ponte Rotto," was seriously damaged and isolated by a flood on the Tiber in 1598.

The 1598 flood crest was lower than the modern embankments of the river (18 to 19 meters [59 to 62 feet]). The walls and embankments along the Tiber were constructed after the unification of Italy and the last catastrophic flood of December 1870. The embankments have provided protection for Rome from most serious flooding. However, deleterious effects come along with modern flood control and diminution of the turbid waters. The effects include the consequences of accelerated riverbed erosion in the controlled channels through the city, decreased sedimentation at the river mouth, and strong erosion of the beaches near Ostia caused by a lack of sand previously brought to the seashore by the Tiber.

Construction of the walls eliminated riverside scenery, cut off narrow country roads, eliminated the stairs of Ripetta and views of the Tor di Nona and Trastevere, and caused damage to the historical riverside palaces like the Palazzo Corsini and the Villa Farnesina. The history of Roman floods and Imperial planning was evident in the distribution of the "urban" marshes along the right and left banks of the Tiber. For example, in Trastevere during the Imperial period, rather than being developed for residences, a marshy area was developed as a venue for mock sea battles. Public facilities such as this great pool were easy to clean after floods, and most residences were not affected by flooding. The medieval urban layout was just the opposite, with massive residential development on the alluvial plain between the two papal residences of the Quirinal and the Vatican; the experience of floods did not serve to decrease the hazard to the poor population stuck on the floodplain.

The Piazza Bocca della Verità was a landing stage in ancient Rome. The piazza is located at the confluence of several small streams that once entered the Tiber (thus, the "V" shape of the piazza). It was the location of the ancient Forum Boarium, a marketplace for livestock as well as a river landing. The importance of this market is evident, for within a radius of several hundred meters you can find the Temple of Portunus, the so-called Temple of Fortuna Virilis (a god who protected the river landing), the so-called Temple of Vesta, or, better yet, the round Temple of Hercules the Conqueror (the most ancient building in the city constructed of marble), and the Arch of Janus's covered passage from the Velabrum to the Forum.

The open ground of the Forum Boarium once was at the confluence of the Tiber and the Velabrum, a perennial watercourse up to medieval time; sometimes when the creek was in full spate, it was necessary to ferry people between the Capitoline and the Palatine. The bank of the Velabrum was where the mythical Romulus and Remus were raised by the she-wolf, and the site was somewhere between the Palatine Bridge and the outlet of the Cloaca Maxima (2nd century B.C.).

The Arch of Janus was constructed in the ancient Forum Boarium, a marketplace for livestock, as well as a commercial landing on the Tiber River. It was modified in medieval times to be part of a fortified home but was restored to its original condition in 1827. The marketplace stood at the confluence of the Tiber and the Velabrum, a major tributary.

The Tiber Island is located at the confluence of the ancient Arenula Creek and the Tiber River at the Forum Boarium. These watercourses, along with the creeks of the Velabrum and Fosso dell'Acque Mariana, came together where the Tiber's floodplain is narrowed by bedrock on the right (the Janiculum) and left (the Capitoline). The island apparently developed rapidly during large floods. Splitting the Tiber into two channels resulted in a more stable channel than that which had existed before.

Stop 5. The Aventine (Aventino) Hill

The most interesting and accessible viewpoint for the Aventine Hill is from the Circus Maximus, reaching Piazzale Ugo de Malfa where streets radiate to the west across the Aventine. This location is also a great

The Tiber Island, a boat-shaped island that has existed since the time of Imperial Rome. At this spot, here viewed from the northwest, the Tiber has split into two channels. This division was natural and not man-made.

place to view the Circus Maximus, the Palatine Hill, and segments of the historical center of Rome. It is also possible, on clear days, to see the volcanic field of the Alban Hills, the source of most of the pyroclastic flow deposits that underlie the seven hills of Rome.

The Aventine Hill is larger and a little more complex geologically than the Palatine Hill, but in some respects it mirrors the latter hill's form and is separated from it only by the small valley that contains the Circus Maximus. The southernmost of the seven hills of Rome and close to the Tiber, it stood in contrast with the Palatine by being home for middle-class citizens. As was the case for the Palatine Hill, the Aventine can easily be seen to be a plateau from all sides. From the Circus Maximus, there is a relatively steep climb to the hilltop, and from the busy Via Marmorata, on the west side of the plateau, it has near-vertical slopes.

The Aventine is the most isolated of Rome's seven hills. The northern slopes, adjacent to the Circus Maximus, consist in part of a sandy

Looking northwest along the axis of the Circus Maximus toward the Tower of Moletta. The Murcia valley became one of the largest sporting venues in the world, with a capacity for 300,000 spectators. The 12th-century A.D. tower was built to defend a mill. (In common with the second field trip itinerary)

claystone, which made up a low hill that was later buried by pyroclastic flows from the Alban Hills. The main part of the Aventine plateau is composed of tuffs, which filled and leveled the older terrain. Romans quarried the tuff underground for use as building stones.

Stop 6. Circus Maximus (Circo Massimo)

The Circus Maximus was established in the Murcia valley along the myrtle-lined stream channel of the Acqua Mariana (in the popular idiom, Marrana) that descended from what is now San Giovanni in Laterano and past Porta Metronia, toward a confluence with the Tiber between the Palatine and the Aventine hills. The ancient stream is now overlain by the Via dell'Amba Aradam, the Via delle Terme di Caracalla, and the Circus Maximus. The tuff plateau that rose above the Valle Murcia contained some small springs along the north side, which served as water sources, along with wells dug in the shallow aquifers along the riverbed. The notable elevation of the tuff plateau and at least the three principal perennial watercourses made it favorable for the development of the first human settlements in Rome; ancient bronze artifacts have been found recently, evidence for the presence of these settlements. The contrasting landforms of plateaus and creek valleys favored the birth and development of the ancestral city, which has long since been radically modified by progressive filling of valleys with debris. The original stream valley was progressively modified and is now buried by the floor and embankments of the Circus Maximus (dimensions: 600 by 200 meters). From an early date, the valley was destined to become a place for spectacles, with a maximum capacity of 300,000 spectators.

The circus was restored by Julius Caesar. Augustus had a 13th- or 12th-century B.C. Egyptian obelisk placed in the circus (that obelisk is now in the Piazza del Popolo). Constantius II brought a 15th-century B.C. obelisk to the circus in A.D. 357 (the oldest Egyptian obelisk in Rome, from the Temple of Ammon at Thebes), which is now in the Piazza San Giovanni in Laterano. Fifty-six Egyptian obelisks were eventually brought to Rome, moved and repositioned from their original locations, with the possible exception of the obelisk in the Circus of Gaius and Nero, which is now the site of Saint Peter's Basilica. In the eastern-center part of the Circus Maximus is the Tower of Moletta, constructed in the 12th century to defend a mill.

Along the central axis of the Circus Maximus was the location of the pope's gasometer built about A.D. 1860 to provide gaslights during a time of public works construction by Pope Pius IX (see chapter 9). Warehouses for bronze, southwest of the Palatine Hill, were located along the north bank of the channel of the Mariana Creek; remnants lie along the Circus Maximus against the Aventine Hill.

Northeast of the Circus Maximus, at the intersection of the Via di San Gregorio and the Via dei Cerchi, you find yourself at the southern edge of the Palatine, with tuffaceous slopes delimiting the southwestern (Cermalus) part of the hill. The Palatine has a quadrangular shape with

Well-preserved ruins of a palace along the southwestern base of the tuff deposits that make up the Palatine Hill.

an elevation of 46 to 50 meters (151 to 164 feet) above sea level. There were springs on three sides of the hill. Most of the hill consists of tuff with optimal geotechnical characteristics for construction materials and site stability. Its southeastern summit may have been a logical site for early settlers.

Stop 7: Celian (Celio) Hill

The Celian Hill is due south of the Colosseum and can be approached along the Via Claudia or the Viale del Parco del Celio (which is also a stop on a tramline, trams 13 and 30b). Although obviously a prominent plateau and important in Roman history, the Celian Hill is least likely to be found in a tourist guidebook. The northern edge of this plateau still retains ruins of the Temple of Claudius, the Churches of San Gregorio Magno and Santa Maria in Dominica, and several very pleasant parks.

As is the case for the other six hills, the Celian Hill has been etched out by erosion of tuffs left by eruptions of the volcanoes of the Alban Hills, southeast of Rome. Man and his works have thoroughly buried and disguised the rocks, but they are evident in their constant use in construction and in boreholes drilled by engineers and geologists. These hills also offer stable platforms for construction.

Stop 8: The Esquiline (Esquilino) Hill

To reach the Esquiline Hill from the south, climb the steps across the street from the Colosseum, which lead to a park on the Oppian Hill, a spur of the Esquiline. From the north, take the Via Merulana southeast a short distance to the Via del Monte Oppio, which leads into the park above the Colosseum.

The Esquiline Hill forms the prominent ridge north-northeast of the Colosseum, up and beyond the traffic booming by on the Via dei Fori Imperiali. Climbing up steps into the green park of the Oppian Hill, you are reaching one of the two most notable surfaces of this plateau. The other, more northeasterly summit is occupied by the great Church of Santa Maria Maggiore.

Tuffs that make up this plateau are overlain in part by a thin veneer of ancient river sediment (gravel and sand). The southern rim of the Esquiline Hill overlooks the Colosseum and a creek bottom partly filled by Nero for his palace. This creek separated the Esquiline Hill from the Celian Hill, which is south of the Colosseum. Ruins now visible on the Oppian Hill include the Domus Aurea, where Nero began rebuilding Rome with his "Golden House" after the great fire of A.D. 64. Much of the house was demolished after Nero died, and the area became the site of Trajan's Baths in A.D. 104–9. The Domus Aurea can be visited after making a reservation.

Stop 9: The Colosseum

The view here at the valley of the Colosseum from the terrace of the Fagutal (located north of the Colosseum, above the Colosseo Metro

The valley of the Colosseum, from the Celian Hill. This valley once accommodated a creek, then Nero's pond, and eventually the Colosseum. The better-preserved northern half of the stadium sits on rock, whereas the damaged southern half overlies poorly consolidated sediments.

Station) illustrates the structure of the Esquiline-Oppian-Fagutal and evolution of the site during prehistoric times, Nero's eventful reign, and eventually the construction of the Colosseum.

Stop 10: The Palatine (Palatino) Hill

Access to the Palatine Hill is by ticket, from the Roman Forum. Follow signs to the various access points: one is via the Clivus Palatinus Road; there is also an entrance on the eastern side along the Via di San Gregorio.

The Palatine Hill is easily identified from all sides, with its prominent tablelike form, covered with ruins and groves of trees. One of Rome's top attractions, the Palatine is believed to be the first of Rome's seven hills to be inhabited and perhaps the original nucleus from which the great city evolved. To get a good feel for the overall form of this small plateau, start at the northwest end of the Circus Maximus, walking southeast. This allows a good view of the end of this rectangular plateau and its ruins. When you reach the end of the Circus Maximus, turn left up the Via di San Gregorio and walk along the eastern margin of the hill toward the Colosseum. This will take you to the entrance to the Roman Forum and your access to the Palatine Hill.

The Palatine Hill is composed of rocks very similar to the sequence that underlies the Capitoline—a series of pyroclastic flow deposits from the Alban Hills that were deposited on a surface of river and marsh deposits. Overlying the tuffs is a veneer of sediment from younger rivers and adjacent marshes. The more or less tabular stack of deposits was later incised by streams flowing from the plateau into the Tiber, leaving the present plateau as an erosional remnant.

Stop 11. The Via dei Fori Imperiali

Benito Mussolini ordered that a broad triumphal avenue be constructed between Piazza Venezia and the Colosseum, cutting through large sections of the Roman fora and medieval neighborhoods. There is evidence here that there was once a connecting low ridge between the

"Scenic wall" along the Via dei Fori Imperiali. These walls were constructed by order of Benito Mussolini to hide poor tenements along his Via dei Fori Imperiali, a wide street linking the Piazza Venezia with the Colosseum. The street was to be used as a grandiose showplace for Fascism.

Quirinal and Capitoline Hills, but that has been destroyed by several thousand years of modification during growth of the city.

The Via dei Fori Imperiali is flanked by "scenic walls" that were erected by Mussolini to hide poor neighborhoods from his triumphal avenue. The walls still stand. Coincidentally, Apollodorus of Damascus, who designed Trajan's Forum and marketplace, and the architect Antonio Muñoz, who in 1934 designed the scenic wall for Mussolini, both decided to follow the break between the tuff plateau and the floodplain.

Cross the Forum by the entrance from the Via dei Fori Imperiali and leave by the exit to the Via della Consolazione.

Stop 12. Via della Consolazione

Along the Via della Consolazione there are outcrops of tuffs typical of those that underlie the seven hills of Rome. Go up Via Monte Tarpeo for an overview of the fora.

Tuff outcrops along the Via della Consolazione, a rare sight in a city covered with buildings and streets. These are the thin distal edges of pyroclastic flow deposits from the Alban Hills volcanic field.

There is a superb panorama of the fora from the eastern edge of the Capitoline Hill. Continue uphill to the summit of the Capitoline.

Stop 13. The Capitoline (Campidoglio) Hill

The Capitoline Hill is one of the most-photographed hills in the world, although most of the tourists with their cameras don't realize the significance of this promontory that forms the western backdrop for the Roman Forum. This small plateau was the center of power and religion for what was Western civilization 2,000 years ago. The Tarpeian Rock, a cliff on the southern edge, was used for throwing traitors to their death. Today the summit contains the Piazza del Campidoglio, designed by Michelangelo, and buildings on three sides, including the Palazzo Senatorio (Rome's Town Hall), the Palazzo dei Conservatori, and the Palazzo Nuovo (the last two are now museums). The northern slope is now occupied by the Vittorio Emanuele Monument, which was built between 1885 and 1911.

The Capitoline Hill is one of the smallest of Rome's seven hills and consists of three main geologic units: (1) the base consists of sediments from the floodplain of the ancient Tiber, including sands and lake deposits, with some interbedded travertine deposited by springs; (2) several different pyroclastic flow deposits (ignimbrites), all erupted from sources in the Alban Hills; and (3) a carapace of river and lake sediments. As with the other six hills, the main framework that holds up this promontory consists of pyroclastic deposits (tuffs). The Capitoline Hill is one of three of the seven hills of Rome close to the Tiber.

Walk back down toward the Via dei Fori Imperiali, past Trajan's Column and up the Via Nazionale to the Via A. De Pretis. Then turn right for half a block to the Piazza del Viminale.

Stop 14. The Viminal (Viminale) Hill

Although a plateau, the Viminal Hill is less evident because of its covering of closely spaced buildings and narrow streets. The relatively flat summit is flanked by northeast-trending low areas that were once creeks and are now the Via Nazionale (northwest of the Viminal) and the Via Cavour (southeast of the Viminal); note that both streets rise toward "headwaters" in the northeast, near the Piazza della Repubblica. The most easily recognized summit of the Viminal Hill is the large edifice of the Ministero dell'Interno, best seen from the Piazza del Viminale at its north-northeast entrance on the Via A. De Pretis. Another familiar public building on the Hill is Rome's Opera House.

Go back downhill along Via A. De Pretis to the northwest. This street changes names, to the Via delle Quattro Fontane. Walk as far as the Via del Quirinale, turn left, and go two long blocks to the Piazza de Quirinale. Along the Via del Quirinale are two parks and Bernini's exquisite Sant'Andrea al Quirinale, the "pearl of the baroque."

Stop 15: The Quirinal (Quirinale) Hill

The Quirinal Hill was, in Imperial Roman times, a residential area. Today the summit is the site of the Palazzo del Quirinale, which is the "official" residence of the president of Italy. The tuffs of the Quirinal Hill overlie deposits at the base of the hill that consist of sand and gravel deposited by the ancient Tiber River. The Palazzo del Quirinale dominates central Rome and has a good view across the city to the east.

From the Piazza del Quirinale go west down the steps to the Via della Dataria, then right (north) along the Via San Vicenzo to the Trevi Fountain. Relax, have a gelato and an espresso—you have visited the seven hills of Rome!

PANORAMAS, PIAZZAS, AND PLATEAUS

It is suggested that you follow this itinerary by foot for the first part of the field trip, from the historical center of Rome (from Trinità dei Monti to the Capitoline), then proceed with motor transportation (private auto, bus, or an Acotral train on the suburban network). If you decide to follow the itinerary with your own auto, you should try to arrive early at Piazza di Porta Capena, where it is possible to park; then, using the Metropolitana (line B, change at Termini, take line A), reach the Piazza di Spagna and ascend the Spanish Steps to Trinità dei Monti.

Stop 1: Trinità dei Monti

The itinerary begins on the panoramic terrace of Trinità dei Monti in front of the church of the same name that was built on ruins of the ancient Villa of Lucullus. In ancient Roman times, the whole area of the Pincian Hill and the Trinità dei Monti was a suburban zone occupied by large villas. The Villa of Lucullus, also called the Gardens of Lucullus, was built around 60 B.C.

View of Trinità dei Monti from the Piazza di Spagna.

The Villa of Lucullus may have been near a point where the Aqua Virgo emerged from its underground aqueduct and crossed into an arch across the Campus Martius. The villa was on the summit of a hill from which, even today, it is possible to admire the urban development of a city in harmony with the original shape of the land. Notwithstanding the intense urbanization, it is possible to see the original topography, including the valley floor of the Tiber, which is underlain by the river's alluvial deposits, the modest relief of the seven hills on the left bank, and the Janiculum–Vatican–Monte Mario hills along the right bank.

The Pincian Hill is mostly consolidated volcanic ash, which was deposited during volcanic activity of the Sabatini volcanic field, north of Rome, and of the Alban Hills volcanic field, south of the city. Interbedded with the tuffs are travertine spring deposits. The tuffs overlie clayey sediments of Pliocene age, now masked by a large staircase (the Spanish Steps). Looking west across palaces that line the Via del Corso (the ancient Via Lata), it is possible to recognize the Tiber's wide alluvial valley, with the river's bend that nearly touches the Castel Sant'Angelo. From Saint Peter's Basilica the land rises toward the Janiculum Hill. Toward the south your view is blocked by the many buildings of the historic center, but on less hazy days you can recognize the monumental edifices capping the seven ancient hills, and the monument to the unknown soldier stepping up the Capitoline's slopes.

From Trinità dei Monti, descend the Spanish Steps to the Piazza di Spagna and walk along the famous Via Condotti, which was most likely the former route leading to the Villa of Lucullus. When you reach the Via del Corso, turn left and walk to the Piazza Colonna (around ten minutes).

Stop 2. Piazza Colonna

The piazza is named for the Column of Marcus Aurelius Antoninus Pius, erected around A.D. 180. The column, which was modeled after Trajan's Column (which we come to at a later stop), rests on a foundation 3.86 meters (12.5 feet) below the present ground level in a pit that was higher than the level of the ancient Via Flaminia. The column is

The Column of Marcus Aurelius, built around A.D. 180. Constructed on poorly consolidated alluvial deposits, the column has been damaged by earthquakes, especially the one occurring in A.D. 1349.

29.6 meters (100 feet) high. On it are carved reliefs describing the war of Marcus Aurelius against the Germans and the Sarmati (tribes living along the Danube). The column has survived centuries of damage and has required considerable restoration. The most evident damage, that of fracturing of the marble cylinders that make up the column, with a dislocation of about 10 centimeters (4 inches) at the tenth cylinder, was caused by an earthquake in the Apennines that shook the City of Rome. It is believed that the damage occurred during the earthquake of A.D. 1349, which had an epicenter near Cassino and Isernia and was the most disastrous historical earthquake to affect Rome. Documents from 1350 describe damage to the column's internal stairs, which made them unusable by visitors who wanted to reach the viewpoint above medieval Rome. From an analysis of earthquake catalogs, it is possible to deduce that at least once every two centuries earthquakes have damaged the celebrated monuments of Imperial, medieval, Renaissance, and baroque Rome.

The most severe earthquake damage has been to buildings constructed on the Tiber's alluvial deposits. A significant example of the importance of underlying geology to structural stability during earthquakes is the contrast between the Column of Marcus Aurelius and Trajan's Column, located only 700 meters (2,300 feet) apart, which were constructed of the same materials and with the same techniques. The Column of Marcus Aurelius is underlain by about 60 meters (197 feet) of recent alluvial deposits. Trajan's Column is located along the valley margins, with its foundation seated in sandstone (bedrock) at the base of the Quirinal Hill.

Leaving the Piazza Colonna, proceed along the Via del Corso toward the Piazza Venezia, passing through the Piazza di San Marcello, then going right along the Via di Piè di Marmo and on through the Piazza della Pigna until you reach the Via dell'Arco della Ciambella, where we see a remnant of one of the distribution basins linking the Aqua Virgo to the Baths of Agrippa (about fifteen minutes).

Proceed by the Via dei Cestari until you reach the Via del Plebiscito, turn left, and proceed toward the Piazza Venezia. Cross the piazza and go for a few hundred meters toward the Via dei Fori Imperiali until you reach the area of Trajan's Forum (around ten minutes).

View across the Forum of Trajan to Trajan's Markets. The markets were built in excavations in the lower slopes of the Quirinal Hill, overlooking the low, swampy area that had been drained by the Cloaca Maxima storm sewer and became the Forum.

Stop 3. Forum of Trajan

The Forum of Trajan was built during the first century by Apollodorus of Damascus. The location of Trajan's Markets was planned to serve the nearby Forum. The markets required a large area, which was excavated in the sandstone and tuffs of the Quirinal Hill's lower slopes. The excavation depth is recorded by the height of Trajan's Column, located near the center of the Forum. This urban development followed closely the original shape of the terrain, and even today, if you stand near Trajan's Column, you can mentally reconstruct the panorama of that time, with the area of the forum across the street (Via dei Fori Imperiali) encompassed by the Capitoline and Aventine hills.

The reclaimed swampy area was enclosed by the lower slopes of the Quirinal Hill and Capitoline Hill to the north and by the Palatine Hill to the south. Before the excavation of the Quirinal Hill's western slope to construct Trajan's Markets, the link between the Quirinal and Capitoline hills was narrower than it is today. Trajan's Forum and the markets were separated by a wall, still visible, that was constructed with blocks of a tuff named *peperino* of Marino. Utilization of the volcanic deposits cropping out in the area of Rome for construction stone was necessary to keep up with urban development. During the first periods of the city's evolution (Archaic period), the most commonly used stone in the state, the so-called cappellaccio, was quarried in tuffs deposited during prehistoric eruptions of the Tuscolano-Artemisio volcano in the Alban Hills and cropping out very close to the heart of the city.

"Cappellaccio" is the name that quarrymen gave to pyroclastic deposits that are almost exclusively fine volcanic ash in which you can distinguish accretionary lapilli or "pisolites" (spherical layers of fine ash wrapped around a granule and looking like a small onion). The scientific name is *pisolitic tuff*, which is a deposit left during highly explosive eruptions that occurred between 0.6 and 0.5 million years ago of the Tuscolano-Artemisio volcano; these fine ash deposits are at least 10 meters (33 feet) thick in the area now occupied by the historical center of Rome. Some Roman caves (underground quarries) in cappellaccio have been discovered below Termini Station; the stone from these quarries was probably used for construction in the overlying Archaic city.

A last look over Trajan's Forum involves a comparative analysis of Trajan's Column with that of Marcus Aurelius, which we visited earlier. Trajan's Column is somewhat older than that of Marcus Aurelius, having been completed in A.D. 113 to commemorate Trajan's campaigns against the Dacians (an indigenous population that lived in the area that is now Romania). On top of the column was a 4-tonne bronze statue of the emperor that was replaced near the end of the 16th century with a statue of Saint Peter. In the foundation, around 12 meters (40 feet) above the footing, is a room with a small door. On the doorpost is an inscription that records the history of excavations along the base of the Quirinal Hill during the construction of Trajan's Markets and the column. The column's foundation rests on lithified sandstones of lower Pleistocene age. This position, along the valley margin, in contact with competent sediments, is of interest to the City of Rome because only minor oscillations have affected the column during earthquakes.

Trajan's Column, which was completed in A.D. 113 to commemorate Trajan's campaigns against the Dacians (an indigenous population that lived in the area that is now Romania). The column's foundation rests on sandstone, giving it the solid footing required to protect it from earthquakes.

Panorama of the Roman Forum from the Capitoline Hill, looking toward the Temple of Antoninus and Faustina. This forum, closest to the Capitoline, developed over many centuries in a topographically low area that was originally a swamp. To drain the swamps and open land for development, a large storm sewer (the Cloaca Maxima) was built and still functions today.

Leave Trajan's Forum, cross the Via dei Fori Imperiali, and walk toward the Capitoline along the Via di San Pietro in Carcere (immediately east of the Vittorio Emanuele Monument). Reaching the Piazza del Campidoglio, take a short walk along Via del Campidoglio and stop at the overlook for a panoramic view of the Roman Forum (around five minutes).

Stop 4. The Forum

The Roman Forum, at the foot of the Capitoline Hill, is situated in a topographically low area that was originally a swamp. Its temples and other structures were built at various times as the forum developed over many centuries. Perhaps this view is one that can be used to best mentally reconstruct the growth of Rome during Republican times. Most of the residential areas were located higher up on the Palatine, Celian, and Esquiline hills. The Aventine Hill was a little too distant, originally separated from the Palatine by a valley later occupied by the Circus Maximus.

The curving valley, originally very swampy and underlain by alluvium from the Tiber and its tributaries, had mostly public structures such as the temples and courts of the Forum. Consequently, the city developed following the lines of the land. Areas subject to flooding were eventually drained by the Cloaca Maxima, allowing public activities in the creek bottoms.

Descend along the narrow street leaving the southeastern corner of the piazza, which leads to the Tarpeian Rock. After a short stroll along the Via della Consolazione, turn uphill and follow the Via di Monte Caprino (five minutes) to the Tarpeian Rock. If you need a break, there is a quiet park along the Via del Tempio di Giove.

Stop 5. The Tarpeian Rock

This cliff has particular significance for every Roman and represents one of the most interesting geologic monuments to Roman history. According to legend, during the war by Rome against the Sabines (4th to 3rd centuries B.C.), a young girl named Tarpeia, who favored the enemies, unlocked the gates to the city and permitted them to surprise the sleeping Romans. In one version of the story, when the Romans recovered from the situation and discovered the betrayal, they punished the girl by throwing her over the cliff. Since that event, this cliff on the Capitoline Hill has been remembered as the Tarpeian Rock, a traditional site for capital punishment of traitors.

The cliff is the edge of an ignimbrite plateau composed of lithoid or *Lionato* tuff, deeply eroded by a stream; the Capitoline is a very typical Roman landform and one of the "seven hills." The Tarpeian Rock is an erosional remnant of this ignimbrite, which was deposited during eruptions of the Tuscolano-Artemisio volcano about 350,000 years ago.

The eruption that deposited the *Tufo Lionato* was the last eruption of the Tuscolano-Artemisio volcano before its summit collapsed, leaving a large collapse crater, or caldera. The edges of this caldera are easily seen in the Alban Hills. In the area of the Capitoline it is possible to reconstruct the evolution of the landscape following the deposition and erosion of the ignimbrite plateau. The ancient valley that existed below the Capitoline crossed near the Roman Forum. The valley was partly filled with alluvium before the arrival of the first pyroclastic flows now called "pisolitic tuffs"; the alluvium now crops out at the base of the Tarpeian Rock, along the Via della Consolazione.

Return to the Via della Consolazione and follow it to the end of the Via del Foro Romano. Turn right and walk downhill along the Via di San Teodoro, then turn right onto the Via del Velabro (five minutes).

Stop 6. San Giorgio in Velabro

To the left of San Giorgio in Velabro is the mouth of an ancient street that is now closed and may have been the site of the Vicus Iugarius; here is a small monumental door or arch dedicated by the moneychangers (bankers) and the merchants to the emperor Septimius Severus and his family. The door may have led directly to the Forum Boarium, a sheltered marketplace near the Tiber Island. The arch consists of two broad pillars dressed with marble slabs on a travertine base and a marble lintel. The Via del Velabro is named after the Velabrum, a swampy area where legend places the discovery of Romulus and Remus by the she-wolf.

From the Arch of Janus, return to the Via di San Teodoro and proceed to the end of the street at the Via dei Cerchi, which skirts the Circus Maximus (five minutes).

Stop 7. Circus Maximus

It took many centuries to complete construction of the Circus Maximus, which may have been the grandest stadium of all time. It was used mainly for sport (races with two-horsed chariots) and occupies much of the Valle Murcia, a stream valley that passed between the Palatine and Aventine hills. The valley was legendarily reclaimed by

The view southeast along the axis of the Circus Maximus. The Valle Murcia, a stream valley between the Aventine and Palatine hills, was drained and modified to become one of the world's largest sports stadiums.

Tarquinius Priscus, an Etruscan king of Rome. In 329 B.C. painted wooden starting gates, or *carceras*, were built along the northern side of the circus. It was during these years that the stream through the valley was most likely confined to a man-made channel that emptied into the Tiber.

In the center of a wide curve on the Palatine side of the valley was a triumphal arch. In 174 B.C., the *carceres* were replaced by brick buildings, and seven egg-shaped objects were added to the central dividing barrier (the *spina*); the "eggs" were used to count laps during a race. During the time of Agrippa, in 33 B.C., the "eggs" were replaced by seven dolphins with the same function. The circus was eventually embellished with two obelisks from the Roman conquest of Egypt. One of the obelisks is now in the Piazza del Popolo and the second in the Piazza di San Giovanni in Laterano. Dionysius of Halicarnassus wrote that during the Augustan period the circus held 150,000 spectators (most modern sports stadiums hold fewer than 80,000 fans). After the fires of A.D. 36 and 64, the stadium was reconstructed to hold 250,000 spectators. After another fire it was reconstructed yet again by Trajan. Most of the stadium's remains are now buried by several meters of human backfill, and shallow groundwater makes archeological excavation in the circus difficult.

Travel southeast along the axis of the circus to the Piazza di Porta Capena (southeast corner of the Palatine). From the piazza, the next stage of the field trip is best done by car. Go southeast along the Viale delle Terme di Caracalla to the end of the street (not more than five minutes, traffic permitting). For those who want to follow this guide using public transportation, it is suggested that you take one of the ATAC buses that follow the Viale delle Terme di Caracalla to the Piazza Numa Pompilio, where you leave the bus and go by foot to the entrance of the Terme di Caracalla (the Baths of Caracalla). If you walk, be advised that the trip from Piazza di Porta Capena to the baths will take twenty minutes, going by the sidewalks along playing fields and a community rose garden.

Stop 8. Baths of Caracalla (Terme di Caracalla)

The baths are at the base of the "little Aventine hill" some distance from the Porta Capena, in a zone rich in natural resources where there were many medicinal springs. The construction was initiated in A.D. 212, and the baths were opened four years later, with the public entrance through the central building. The baths functioned continuously through the 3rd and 4th centuries A.D. and partially during the two subsequent centuries, when there were notable structural modifications. The baths were probably closed when the aqueduct supplying water was cut by the Goths in A.D. 537 during an assault on Rome.

The baths followed the slope of a hill composed of volcanic deposits left by eruptions in the Alban Hills, which overlie sedimentary rocks, including limy mudstones representing a river and lake environment, which in turn overlie gravelly sandstones deposited in a river delta. The gravelly sandstones host the most important aquifer of central Rome and are the source of many springs. With the construction of the baths the little hill and the area immediately around it were profoundly modified.

The east-facing hill is the site of an imposing man-made terrace with three levels having a total height of 14 meters (46 feet). The highest

The Baths of Caracalla.

terrace coincides with a street and served as a foundation for a water tank that served adjacent buildings. The intermediate level, about 9 meters (30 feet) lower, was the hypogeum, and the lowest level, about 14 meters (46 feet) below the present hilltop, was the foundation for the central building and contained drains for the baths.

From the Baths of Caracalla, return to the Piazza Numa Pompilio, enter the Via di Porta San Sebastiano, and continue to the monument of the Tomba degli Scipioni (the Tomb of the Scipios) (by auto about ten minutes; by foot it is possible to take an ATAC bus from the Piazza Numa Pompilio toward the Porta San Sebastiano and ask the conductor to indicate the stop for the Sepolcro degli Scipioni (two or three stops).

Stop 9. Tomb of the Scipios

This sepulchre was excavated in the pisolitic tuff in what was likely an ancient underground quarry. Parts of the sepulchre are at least as old as the early 3rd century B.C., corresponding to a sarcophagus of that age.

The area of the sepulchre follows outcrops of a sequence of early ignimbrites (pyroclastic flow deposits) from the Tuscolano-Artemisio volcano. The rapid emplacement of pyroclastic flows caused a drastic modification of the countryside, displacing the course of the ancient Tiber River (the route of the ancient Tiber was very different than that of the river today). Geologists have found deep in drill holes near the sepulchre blue clays, indicating that the volcanic sequence overlies clayey sands left by the Tiber during the middle Pleistocene.

Near the southwest corner of the monument, where a path leads to a lower level, you can see outcrops of a brown paleosoil, with a thickness of 30 centimeters (12 inches), which has the characteristics of an *andosol* (a clayey brown soil that develops rapidly on volcanic ash deposits). Overlying the paleosoil is the pisolitic tuff, the lowest portions of which have been altered to clay. Near the *calcara* (the place where, during medieval times, limestone and marble were burnt to make lime for mortar) at the southwestern corner of the sepulchre, the ignimbrite unit contains the remains of trees blown down by the pyroclastic flows.

The top unit is very rich in accretionary lapilli (pisolites); interbedded with this unit of the pisolitic tuff are notable layers, ranging in color from brown to gray, that were left by eruptions of the Sabatini volcanic field north of Rome. Also near the top of this rock sequence is a deposit of coarse yellow-ochre sand. Beginning with the Tomb of the Scipios is a zone occupied by sepulchres of various ages, of which the most interesting are underground columbaria of five rows overlying semicircular niches coated with well-preserved plaster cornices. Each one of these shelters has two terra-cotta coats. The walls have traces of plaster with lively paintings. It has been calculated that the columbarium had more than 470 urns containing human ashes.

From the Tomb of the Scipios, proceed along the Via di San Sebastiano to the gate of the same name, which opens onto the ancient Appian Way and continues on to the Tomb of Caecilia Metella (about fifteen minutes by car or by ATAC bus, which you can pick up near the Porta di San Sebastiano and follow the ancient Appian Way; ask the the conductor to indicate the stop nearest the Tomb of Caecilia Metella). If you have enough time, stop at Porta di San Sebastiano and visit the Museo delle Mura (the Museum of the Walls), which is one of the most overlooked and interesting museums in Rome.

Stop 10. The Tomb of Caecilia Metella

This may be the mausoleum of the most famous Roman of the ancient Appian Way. It is dedicated to Caecilia Metella. Its circular form provides a panoramic view of the Roman countryside and remains a fundamental point of reference, being visible from the surrounding region. The tomb is located at the end of the Capo di Bove (head of the bull) lava flow, which takes its name from a carved bull's head near the door of this funeral monument.

The Capo di Bove lava flow is a geologic element of the countryside that is linked to Roman culture because of its plateaulike form that rises above the surrounding terrain and on which the Romans built the Appian Way, the first consular road to connect the city with villages of the Alban Hills. The Capo di Bove lava flow contains leucite crystals and was erupted from the Faete cone (a volcano rising above the floor of the Tuscolano-Artemisio caldera) about 280,000 years ago. The lava flows filled a valley that extended northwest from the central volcanic field toward Rome, probably to the brim, and stopping near the spot now occupied by the Tomb of Caecilia Metella. The thickness of the lava flow at the tomb is about 10 to 15 meters (33 to 49 feet). Near the tomb are numerous abandoned quarries that were open during Roman times along the lava flow margins, where stone was quarried for paving stones used in the ancient Appian Way and subsidiary roads.

The Capo di Bove lava flow stands out above the countryside because it is several meters higher and resistant to erosion. In fact, the resistant lava flows that once filled the stream valley are now visible as "inverted topography," where the crumbly pyroclastic flow deposits that originally formed the valley walls have since been eroded, leaving the lava filling as a ridge. The ancient Appian Way follows the top of this ridge.

Proceed along the Via Appia Antica (ancient Appian Way) until you encounter the Via di Fioranello. Go around the bend to the left and proceed to the crossing of the new Appian Way and follow it to a sign for the Via dei Laghi. The entrance to this street is located where the terrain begins to rise toward the volcanic caldera margins and where it is possible to see on the right side of the road the numerous vertical walls that remain from ancient Roman quarries excavated in the *pepe-*

The Tomb of Caecilia Metella. This may be the mausoleum of Caecilia Metella. Located at the end of a lava flow from the Alban Hills, the tomb is decorated with the head of an ox. The ox head provided the name used by geologists for the flow, called the "Capo di Bove" lava flow.

Paving stones along the Appian Way. These *basoli* were carved from basaltic lavas.

rino of Marino, where tuff blocks were cut for construction (by auto, about twenty minutes; by public bus, take an extraurban line that goes from the city to the hills along the new Appian Way).

Stop 11. Roman Quarries along the Via dei Laghi

The *peperino* of Marino are tuffs from the final eruption of the Albano crater about 30,000 years ago; most were erupted from the southernmost of the five coalesced craters that now make up a lake basin (next stop).

During the violent eruption in which the magma was strongly fragmented, ash was thrown out of the crater mixed with gases and sedimentary rock fragments broken from the magma chamber walls. The extremely fluid pyroclastic flows moved rapidly from the volcano and filled valleys. Here on the left side of the road, the valley-filling tuffs have total thicknesses of 30 to 35 meters (98 to 115 feet). As the tuffs cooled, the ash particles were cemented by minerals formed when residual steam in the deposits reacted with glass volcanic shards and formed zeolite minerals.

On the wall of the quarry you can see other characteristics of the rocks that appear to be geologic textures but are really marks left by the quarrymen's tools. The gray color of the deposit masks the clear gray of the fine-grained ashy matrix that contains leucite, mica, and small pieces of lava. The *peperino* of Marino was used by the Romans as some of the oldest ornamental stone and as blocks for construction of buildings, walls, and so forth, during Imperial times. Examples of the *peperino* of Marino can be seen in the Forum (the Temple of Antoninus and Faustina) and in Trajan's Forum (the wall that separates the Forum from the markets).

Proceed along the Via dei Laghi to the crossroad for Marino. The village of Marino was constructed along the margins of a valley filled with a deposit of *peperino* of Marino; thus, most of the homes in the village are built with peperino blocks, and the material is also used as an ornamental stone for doorways and monuments. The famous fountain in Marino flows with wine in October to commemorate the battle of Lepanto against the Ottomans. In remembrance, a bas-relief was cut into blocks of *peperino*, representing the figures of imprisoned enemies (about twenty minutes to Marino; for public transportation, it is necessary to take a tram that proceeds toward Velletri or Nemi, and it is not possible to visit Marino).

At the crossroads, after the traffic signal, proceed toward the second set of signs for Velletri and Nemi. After going a few hundred meters to the right, you can see a beautiful panorama of Lake Albano.

Stop 12. Lake Albano

The Albano lake basin is made up of five overlapping craters that were formed during hydrovolcanic eruptions of the last stages of activity in the Alban Hills. Hydrovolcanic eruptions occur when rising magma mixes explosively with water in a regional aquifer that underlies the volcanic field. These eruptions left the five northwest-southeast-trending craters. Each of the coalesced craters can be distinguished by their shapes (easily identified by bathymetric surveys of the lake) and by the diverse rock types that distinguish the craters and the sequence of their development. The craters that make up Albano are in fact explosion craters excavated along the walls of the Faete cone, with the southern craters being topographically higher than the northern ones.

The southern crater margins, cut into the lava flows that form the Faete cone edifice, are resistant to erosion and form the steepest slopes along the lake margins. The northernmost craters were excavated in the area behind the break in slope of the volcano, along the outside zone that consists of ignimbrite plateaus that are less resistant to erosion than the lava flows. These craters therefore have gently sloping walls that are much less steep than those in the north. The Via dei Laghi descends to the lake (currently closed to traffic), along which there are exposures of the Albano eruption sequence.

Inside the crater, at the base of the road near lake level, you can see the lava flows that make up the Faete volcano, which are, in turn, overlain by the peperino that was erupted during the formation of the craters. In some cases the pyroclastic flows that formed peperino deposits did not rise over the crater walls and flowed back toward the crater floor.

Along the Via dei Laghi, follow the signs for Grottaferrata-Frascati. Reaching the village of Frascati, at the main plaza you turn onto the road for Tuscolo (by auto, about thirty minutes; by public transportation, it is necessary to change at the junction for Frascati, cross the village of Frascati, and change to a new bus for Tuscolo).

Stop 13. Tuscolo (Tusculum)

Here, at the final point in our field trip, there is a spectacular panorama where you can see the overall shape of the Alban Hills volcano. Tuscolo is particularly loved by Romans for the feeling of what was typical Roman countryside. Ancient Roman life is reflected in a perfectly preserved small theater excavated in deposits left by lava fountains that erupted along a fissure, an activity that accompanied the collapse of the Tuscolano-Artemisio caldera. The road winds steeply upward and is cut into these deposits of welded scoria (cinders) that form the summit of Tuscolo (a little below the cross), where it is possible to observe the deposits passing laterally and vertically into lava flows.

From the viewpoint at the cross of Tuscolo you can see crater walls that enclose the Tuscolano-Artemisio caldera. The flat bottom of the basin is the Tuscolano-Artemisio volcano caldera floor. By extending upward the truncated volcano slopes, you can envision the shape of the large cone that was there before the crater collapsed. The caldera floor hosts the pine-covered Campi di Annibale, a famous stronghold of soldiers during the war against Rome. Along the rim of the Faete cone you can see the large cone of Monte Cavo, following the border of the eastern edifice during the collapse phase of Faete and which can be recognized by the large cavity that contains the craters of Albano.

From here, retrace your route to Rome or follow the better-traveled roads used by Roman commuters who live in the Alban Hills. After a tiring day of geology and history, stop along the way to sample the white wines of the Castelli Romani and perhaps have a late lunch at one of the trattorias.

View of the Alban Hills volcanic field from Tuscolo.

A FIELD TRIP TO ROME, THE CITY OF WATER

HUMANS and water are inseparable. Since time immemorial, humans have selected places for development based on the ability to find water, including the sea, rivers, lakes, or springs. Water is essential for life and all human activities, such as transportation, commerce, and industrial and technical development. The great cities of history and the growth of both ancient and modern megacities have been linked to water.

Rome is no exception to this rule. In many respects it represents the prototype of cities whose development is linked to the element of water. Rome has abundant water, with sources supplying about 24 cubic meters per second (380,448 gallons per minute). Rome's fortunes have depended on water, including its river, for success. During the first phases of the city's development, the water was not a guarantee of the city's success in that it was not potable, leading to high mortality rates, even though it provided communication and commercial links with the outside world.

This field trip involves a route that reaches historical sites by following the natural life of water for the Romans, from ancient springs that are now buried and other interesting archeological sites to the aqueducts that supported Roman health.

Stop 1. Springs at the Base of the Vatican Hill

This field trip begins at the base of the Vatican Hill and follows the right bank of the Tiber, crossing the Tiber Island and proceeding toward the historical center of Rome. Then it returns to and follows the left bank of the Tiber, reaching more peripheral areas. During this walk, you will see some of the city's springs, and a brief look at the aqueducts that you pass during the walk will help illustrate the story of water use in the city.

Acqua Pia:

- The original location of this spring on the eastern slopes of the Janiculum Hill is today marked by a fountain and pool along the walls of the Vatican near Sant'Uffizio, along the Via delle Fornaci (diagonally across from the Via di Porta Cavalleggeri). This basin was constructed in 1942 to reestablish an older fountain that was destroyed during the construction of the Galleria Amedeo Savoia Aosta. The older fountain was fed by a spring that emerged from an aquifer located between gravelly sand deposits overlying Plio-Pleistocene claystones. The spring was discovered in A.D. 500 and subsequently piped to its present outlet for public use.
- A few meters closer to the Castel Sant'Angelo, after crossing Borgo Pio, you reach the Piazza delle Vaschette with the water of the spring of Acqua Angelica.

Stop 2. The Acqua di Santa Maria delle Grazie, or the Acqua Angelica (Piazza delle Vaschette)

This spring was discovered in 1697 by a friar of the Convent of Santa Maria delle Grazie. Cassio (1756) placed the spring at the base of the slope of the eastern gate of the Teatro Belvedere, where a fountain was

Acqua Pia. This basin was constructed in 1942, replacing an older fountain that was destroyed by construction.

built, diagonally across from the Church of Santa Maria delle Grazie, then moved to the Piazza delle Vaschette when it became part of the Piazza Risorgimento. This is one of the springs of the Vatican Hill that emerged from gravelly sand deposits that overlie impermeable Plio-Pleistocene mudstones.

- From the Piazza delle Vaschette, turn in the direction of the Vatican and go toward the Hospital of Santo Spirito; when in the vicinity of the hospital, go down steps to the bank of the Tiber, where you will encounter a tablet that records the story of another important spring.

Stop 3. Acqua Lancisiana

The Acqua Lancisiana, one of the most notable of Roman springs, was used up to the middle of the 20th century. This spring is famous for the excellent quality of its waters, which are believed to have therapeutic properties. The Acqua Lancisiana rose in the vicinity of the Hospital of Santo Spirito below Sant'Onofrio. A depression is evidence for a tunnel of typical Roman construction that, for 150 meters, followed gravelly sand deposits.

After having been abandoned for many years, the spring was rediscovered by the doctor Lancisi, who, it is said, began using it to cure maladies in the area of the Hospital of Santo Spirito. Pope Clement XI, who wanted to make the waters available to the public, ordered that the spring be relocated and the water be collected in three large pipes to fill a large marble basin upon which was to be mounted an ornamental plaque. Access to the fountain was closed in 1827 to allow work on an extension of the hospital, but vigorous protests by the public induced Pope Pius VII to restore public use. The waters of this spring, boosted by additional collectors, flowed into a new fountain at the Porto Leonino. The fountain was removed at the end of the century because of work being done on walls along the Tiber. The only remaining evidence for the ancient spring, now cut off by the walls, are the two niches from which water flowed.

- The niches are located immediately below two large staircases, along the Tiber at the level of the Ponte Principe Amedeo Aosta, that descend to

Acqua delle Grazie. This is one of the springs of the Vatican Hill that emerge from gravelly sand deposits that overlie older mudstones.

> river level. On the two niches are plaques that commemorate the papal intervention in favor of the fountains.

It is interesting to mention an industrial exploitation of this miraculous water. In 1924 a twenty-year concession was awarded to bottle and sell the waters of the Acqua Lancisiana. To exploit the waters, two channels were run to an industrial park located along the Viale

Acqua Lancisiana. The Lancisiana is reputed to have therapeutic properties and has been moved several times because of construction.

Gianicolo. However, the urban development of the City of Rome polluted the spring waters, which became unusable by 1942 when factories were constructed.

- From the Hospital of Santo Spirito, proceed along Lungotevere to the Via della Lungaretta and enter the Botanical Gardens (which were once the Palazzo Corsini) to visit the variety of springs used in the past to supply water to the Palazzo Corsini.

Stop 4. The Acque Corsiniane and the Botanical Gardens

A hydrant, fed by springwater, captures a source that was originally Roman and probably supplied water to villas in the area. It was rediscovered and restored in 1880 during the construction of walls along the floodplain of the Tiber.

- Leave the Botanical Garden and proceed along the Via della Lungara to where it ends at the Via Garibaldi. Then proceed by foot upslope along the Via di San Pancrazio and the convent of San Pietro in Montorio.

Stop 5. Acqua Innocenziana or the Fountain of the Structure of the Janiculum

This spring was discovered in 1682 during the construction of three mills, ordered by Innocent XI. Cassio (1756) describes their discovery and restoration during the construction of the first mill in the Parnasio Forest, where there are gravelly sandstones overlying claystones.

- At the head of the Via Garibaldi, after the Largo di Porta San Pancrazio, you should proceed along the Via Angelo Masina until you reach the American Academy, in which is an access door that leads into the Aqua Traiana (access permission is required).

Alternate Stop 5. Aqua Traiana

Aqua Traiana was constructed in A.D. 109 primarily to serve the heavily populated quarter of Trastevere, where it was necessary to provide potable water to replace polluted water from the Tiber. Springs feeding the aqueduct are located near Bracciano Lake, north of Rome; the aqueduct reaches the Janiculum Hill and then descends to Porta San Pancrazio.

- Proceed along the Gianicolo wall until you reach the Via Calandrelli and hence the Via Dandolo, where additional springs can be found.

Stop 6. Spring of the Temple of the Syrian Goddess

These waters were used near the end of the Roman epoch, to supply the Temple of the Syrian Goddess.

- From the Via Dandolo, you join the Viale Trastevere, where you meet the Lungotevere (Lungotevere degli Anguillara) and go to the head of the Tiber Island. Use the historic bridges to cross the island, then proceed to the left bank of the Tiber, where you enter the historical district. Proceed on foot to the Campidoglio, along the Via di San Pietro, to reach the Carcer Mamertinus, the ancient city jail.

Stop 7. Tullianum

The Tullianum derives its name from an underground cell in the ancient jail, which contained a water source that flowed continuously from a hole in the floor. The Tullianum fountain is a frequently described historical fountain because its existence is linked to the history of the most ancient and famous of Roman jails, which contained the small cell of the apostle Peter. The legend includes, in fact, an account in which the water spurted out miraculously when Peter needed it to baptize those who wished to become Christians. It is now believed that the Tullianum is the former Fontinali spring.

The spring flow is from a gravelly sand aquifer. The water follows a network of tunnels of the Roman epoch that ultimately drain into the Cloaca Maxima.

- From the Campidoglio go toward the Roman Forum to follow the development of the city, with a brief deviation toward the Piazza Venezia, from which you enter the Via del Corso and finally reach the Via della Merced, crossing an alley to intersect the Via del Nazareno. This little street contains many treasures, including more arches of the Aqueduct of the Aqua Virgo.

Stop 8. The Lacus Juturnae

Rising from the pavement between the Temple of Castor and the Shrine of Juturna at the foot of the northeastern corner of the Palatine Hill is a square marble basin, the Lacus Juturnae. The spring was dedicated to nymph Juturna, and fed a swampy zone between the Palatine and the Capitoline hills flowing toward the Cloaca Maxima. The water came from a line of springs near the Temple of Castor.

- Exit the Roman Forum in the area of the Colosseum and stay on the Via Labicana to reach the foot of the Celian Hill, joining on the right the Piazza di San Clemente. There enter the Church of San Clemente.

Stop 9. Acqua di San Clemente

Underlying the Church of San Clemente, water rises from a Roman pipe. The spring below the church was covered by the Capucin monks above the level of a Republican era house, which in turn overlies an even older structure. The spring flows from gravelly sands.

- From the Piazza di San Clemente go to the Via Celimontana and on to the intersection with the Largo della Sanità militare. Descend by the Via di S. Paolo di Croce, finally reaching the Piazza dei Santi Giovanni e Paolo, and on to the Clivo di Scauro to where it reaches the Piazza di San Gregorio and finally the Piazza di Porta Capena, where a 16th-century building has a plaque that commemorates the spring of Acqua di Mercurio.

Stop 10. Acqua di Mercurio

Located at the base of the Celian Hill between San Gregorio and Villa Mattei, this spring supplied a water lily pond. Today the spring location is recorded on a commemorative plaque on a building on the Piazza di Porta Capena. According to studies conducted in the 19th century, it appears that the waters from this spring flowed under a Capucin building parallel to the valley of the Circus Maximus as far as the Church of Sant'Anastasia. Under the church, even today, significant flow continues through a pipe that probably reaches the Cloaca Maxima.

- Enter the Viale delle Terme di Caracalla where it rejoins the Baths of Caracalla, where a well in the northern gymnasium commemorates the riches of water in this zone.

Stop 11. The Public Baths of Caracalla

In the neighborhood of the Baths of Caracalla there were numerous springs, making the early public baths the most notable in the Valle delle Camene (the Viale delle Terme di Caracalla follows the course of

Acqua di Mercurio. This spring is commemorated by a plaque on this building located on the Piazza di Porta Capena.

the ancient valley). The hydrogeologists Lombardi and Corazza established that Roman sewers ran below the streets from the southeast to the northwest of the baths and carried all the runoff from the Aventine Hill. The water was also used to fill tanks and flush sewers—now visible in the northern gymnasium of the Baths of Caracalla. It is probable that the springs supplying the public baths were cut off during construction and diverted for the purpose of flushing the deep sewer.

The baths were constructed on the slopes of the little Aventine hill some distance from the Porta Capena. Construction was initiated in about A.D. 212, and the baths were inaugurated four years later, with the opening of the public entrance to the central building. The baths functioned without interruption during the 3rd and 4th centuries A.D. and intermittently for several centuries during which there was notable reconstruction and modification. The baths were closed when the aqueducts were cut by the Goths in A.D. 537 during a siege of Rome.

The slopes on which the baths were constructed are on volcanic rocks erupted in the Alban Hills, which buried limy mudstones deposited in a river-lake environment. Interbedded with the limy mudstones are gravelly sandstones of ancient streambeds that now serve as an important aquifer in the Roman area. With construction of the baths the small hill was profoundly modified—for example, to create the monumental street that leads to the baths, the Via Nova.

The eastern slope of the hill is the site of an impressive excavation with a total height of about 14 meters (46 feet) that created terraces at three levels; the highest level coincides with a modern street and served as the foundation for a water tank and adjacent buildings. The intermediate level, about 9 meters (30 feet) lower, held the *hypogaea* (lower levels of the buildings), and the lowest level served as the foundation for the central edifice, including the plumbing necessary to operate the baths.

- Move uphill from the baths toward San Giovanni to the Piazza della Navicella, where there are the great arches of another important Roman aqueduct, that of the *Arcus Neroniani*, a branch of the Aqua Claudia built under Nero.

Alternate Stop 11. Aqua Claudia

According to Fontinus, construction of the Aqua Claudia was begun by Caligula in A.D. 38 and finished by Claudius in A.D. 52. The water came

Aqua Claudia. Aqueduct remnants can be found all over Rome and are best seen in the Aqueduct Park (Parco Appio Claudio), located between the Via Appia Nuova and the Via Tuscolana.

from springs located at Mile 38 of the ancient Via Sublacensis, which ran between Rome and Subiaco, and was carried to Rome for 70 kilometers (43.5 miles); 15 kilometers (9.3 miles) of the aqueduct were above ground. The aqueduct reaches Rome where there are channels from two aqueducts, one of them the Claudian, stacked on top of Porta Maggiore.

This ends your walking tour to see aspects of the infrastructure that supplied water to Rome through history. Visiting sites of springs in the volcanic fields or in the Apennines will require a car and is beyond the scope of this field trip.

Acknowledgments

THE LIST of those who helped with the support of this project is long. Wes Myers and James Aldrich, of the Los Alamos National Laboratory, made the six-month leave possible. The Council for International Exchange of Scholars (Fulbright Scholar Program) provided generous support. Luigi Filadoro of the Fulbright Foundation's Rome office was especially helpful.

At the University of Roma Tre, many faculty, students, and office staff made work on this project possible and pleasant. Many shared their expertise and experience, and sometimes their apartments—Francesca Cifelli, Nicola D'Agostino, Claudio Faccenna, Francesca Funiciello, Ciro Gianpaolo, Guido Giordano, Maurizio Parotto, and Antonio Praturlon. Experts throughout the city of Rome also shared their expertise. Thank you all!

Elsewhere in the Roman community, Heiken thanks Peter and Cynthia Rockwell for sharing their expertise on stoneworking (and a course on stone carving) and monument preservation, their knowledge of Rome, and their friendship.

Most of the superb illustrations in this book are by Daniela Riposati of the Italian National Institute for Geophysics and Volcanology. She was very patient throughout the many changes that occurred during the preparation of *Seven Hills.* Photographs are by the authors and Alessandra Esposito of the University of Roma Tre. The manuscript was ably reviewed by Ian MacGregor and Jill Andrews in the United States and by faculty at the University of Roma Tre and research staff at the National Institute for Geophysics and Volcanology.

Princeton University Press has been capably represented by Kristen Gager and Jack Repcheck, but the main editing and advice have been provided by Joe Wisnovsky, Sarah Green, Hanne Winarsky, Susan Ecklund, and Linny Schenck.

Last, but certainly not least, Heiken thanks his wife, Jody, who has supported him throughout the project and countless lost evenings and weekends.

Further Reading

CHAPTER 1. INTRODUCTION TO THE GEOLOGY OF ROME

Bigi, G., Cosentino, D., Parotto, M., Sartori, R., and Scandone, P. 1983. *Synthetic Structural-Kinematic Map of Italy.* Scale 1:2,000,000. Consiglo Nazionale delle Richerche, Progetto Finalizzato Geodinamica, Florence. (Part of the boxed set of maps—Structural Model of Italy—which, if you are a geologist, are very valuable and represent a phenomenal piece of work by the geologists of Italy.)

Cardilli, L. (ed.). 1991. *Fontana di Trevi—La Storia, Il Restauro.* Edizioni Carte Segrete, Rome. (Contains the analysis by Peter Rockwell.)

Cipollari, P., and Cosentino, D. 1992. Considerazioni sulla strutturazione dei Monti Aurunci: Vincoli stratigrafici. *Studi Geolici Camerti,* special volume 1991/2 CROP 11: 151–56.

Cosentino, D., Parotto, M., and Praturlon, A. 1993. *Guide Geologiche Regionali—14 Itinerari-Lazio.* Società Geologica Italiana, Rome. (This excellent field guide to Lazio is one of many published for Italy by the Italian Geological Society. Even if you don't read Italian, it is worth the reasonable price for the maps and illustrations.)

Faccenna, C., Funiciello, R., and Marra, F. 1995. Structural Framework of the Geology of Rome. In R. Funiciello (ed.), *La Geologia di Roma—Il Centro Storico,* pp. 31–47. Vol. 50 of Memorie descrittive della Carta Geologica d'Italia. Ist. Poligrafico e Zecca dello Stato, Rome.

Faccenna, C., Mattei, M., Funiciello, R., and Jolivet, L. 1997. Styles of Back-Arc Extension in the Central Mediterranean. *Terra Nova,* 9: 126–30.

Hsu, K. J. 1983. *The Mediterranean Was a Desert.* Princeton University Press, Princeton.

Morton, H. V. 1966. *The Fountains of Rome.* Macmillan, New York.

Parotto, M., Salvini, F., and Tozzi, M. 1996. Geologia di Superficie e Geometrie Profonde nell'Italia Centrale: per un Profilo di Previsione CROP 11 da Civitavecchia a Vasto. *Memorie Società Geologia Italiana,* 51: 63–70.

Pinto, J. A. 1986. *The Trevi Fountain.* Yale University Press, New Haven.

Talbert, R. J. A. (ed.). 2000. *Barrington Atlas of the Greek and Roman World.* Princeton University Press, Princeton.

Chapter 2. The Capitoline Hill

Trigila, R. (ed.). 1995. *The Volcano of the Alban Hills.* S.G.S., Rome.

Chapter 3. The Palatine Hill

Charola, A. E. (ed.). 1994. *Lavas and Volcanic Tuffs.* International Centre for the Study of the Preservation and Restoration of Cultural Property, Rome.

De Rita, D., Lucchitta, I., and Giampaolo, C. In preparation. Rome, the City That Owes Its Success to Tuffs. Accepted for the *Tuff Handbook* (ed. G. Heiken).

Fisher, R. V., Heiken, G., and Hulen, J. 1997. *Volcanoes—Crucibles of Change.* Princeton University Press, Princeton.

Marta, R. 1986. *Tecnica Costruttiva a Roma nel Medievo* (*Construction Techniques of the Middle Ages of Rome*). In both Italian and English. Edizioni Kappa, Rome.

Pergola, Phillippe. 1989. *Catacombe di Roma e Necropoli Vaticana.* Vision, S.R.L., Rome.

Chapter 4. The Aventine Hill

Gisotti, G., and Zarlenga, F. 1998. La geologia della città di Roma tra urbanistica e archeologia. *Geologia dell'Ambiente,* no. 4, 48–58.

Pavia, Carlo. 1998. *Guida di Roma Sotterranea—Gli Ambiente Più Suggestivi del Sottosuolo Romano.* Gangemi Editore, Rome.

Società Italiania di Geologia Ambientale. 1998. Geologia Urbana. Special issue of *Geologia dell'Ambiente,* no. 4.

Chapter 5. The Tiber Floodplain

Bellotti, P., Chiocchi, F. L., Milli, S., Tortora, P., and Valeri, P. 1994. Sequence Stratigraphy and Depositional Setting of the Tiber Delta: Integration of High-Resolution Seismics, Well Logs, and Archaeological Data. *Journal of Sedimentary Research* B64: 418–32.

Bencivenga, M., Di Loreto, E., and Liperi, L. 1995. Il regime idrologico del Tevere, con particolare riguardo alle piene nella città di Roma. In R. Funiciello (ed.), *La Geologia di Roma*—Il *Centro Storico,* pp. 125–72. Vol. 50 of

Memorie descrittive della Carta Geologica d'Italia. Ist. Poligrafico e Zecca dello Stato. Rome.

Bono, P., Malatesta, A., and Zarlenga, F. 1995. Itinerario No. 3—Da Velletri a San Felice Circeo (km 150). In A. Praturlon (ed.), *Lazio—Guide Geologiche Regionali*, pp. 117–30. Società Geologica Italiania, Rome.

Braudel, F. 1966. *The Mediterranean and the Mediterranean World in the Age of Philip II.* Vol. 1. Harper Colophon, New York.

Lamb, H. H. 1995. *Climate, History, and the Modern World.* Routledge, London.

Morolli, G. 1982. I progetti di Garibaldi per il Tevere. In *Garibaldi. Arte e Storia, Museo del Palazzo di Venezia*, pp. 93–112. Centro Di, Rome.

Potter, T. W. 1987. *Roman Italy.* University of California Press, Berkeley.

Chapter 6. The Tiber's Tributaries

Amanti, M., Gisotti, G., and Pecci, M. 1995. I dissesti di Roma. In R. Funiciello (ed.), *La Geologia di Roma—Il Centro Storico*, pp. 219–48. Vol. 50 of Memorie descrittive della Carta Geologica d'Italia. Ist. Poligrafico e Zecca dello Stato, Rome.

Barker, G. 1981. *Landscape and Society—Prehistoric Central Italy.* Academic Press, London.

Boschi, E., Guidoboni, E., Ferrari, G., Valensise, G., and Gasperini, P. 1997. *Catalogo dei forti terremoti in Italia dal 461 a.C. al 1990.* Istituto Nazionale di Geofisica, Rome.

Cafiero, G., and Conte, G. 1995. *Roma Città Sostenibile* (Rome as a Sustainable City). Comune di Roma and Ecomed, Rome.

Camassi, R., and Stucchi, M. 1997. *NT4.1.1, un catalogo parametrico di terremoti di area italiana al di sopra della soglia del danno*, GNDT, Milano. Available at: http://emidius.itim.mi.cnr.it/NT/home.html.

Donati, S., Funiciello, R., and Rovelli, A. 1998. Seismic Response in Archeological Areas: The Case-Histories of Rome. *Journal of Applied Geophysics* 41: 229–39.

Farina, G. (ed.). 1985. *Palazzo Valentini.* Editalia, Rome.

Filippi, G. D. 1989. *Guide del Vaticano.* Fratelli Palombi Editore, Rome.

Funiciello, R., and Leschiutta, I. 1993. Carateri geologici e risposts sismica: Il caso di Roma e I suoi monumenti. *Geo-Archeologia* 1993–91: 83–97.

Marra, F., and Rosa, C. 1995. Stratigrafia e assetto geologico dell'area Romana. In R. Funiciello (ed.), *Geologia di Roma—Il Centro Storico*, pp. 49–118. Vol. 50 of Memorie descrittive della Carta Geologica D'Italia.

Molin, D., Castenetto, S., Di Loreto, E., Guidoboni, E., Liperi, L., Narcisi, B., Paciello, A., Riguzzi, F., Rossi, A., Tertulliani, A., and Traina, G. 1995. Sismicità di Roma. In R. Funiciello (ed.), *La Geologia di Roma—Il Centro Storico*, pp. 331–408. Vol. 50 of Memorie descrittive della Carta Geologica d'Italia.
Molin, D., and Guidoboni, E. 1989. Effetto fonti effetto monumenti di Roma. I terremoti dall'anticha ad oggi. In E. Guidoboni (ed.), *I Terremoti prima del Mille in Italia e nell'Area Mediterranea*, pp. 194–223. ING, Bologna.
Provincia di Roma. 1997. *Relazione Sullo Stato dell'Ambiente a Roma—Anno 1997*. Maggioli Editore, Rome.
Rodriguez-Almeida, E. 1984. *Il Monte Testaccio—Ambiente, Storia, Materiali.* Edizioni Quasar, Rome.

Chapter 7. The Western Heights

Amanti, M., Gisotti, G., and Pecci, M. 1995. I disessti di Roma. In R. Funiciello (ed.), *La Geologia di Roma—Il Centro Storico*, pp. 219–48. Vol. 50 of Memorie descrittive della Carta Geologica d'Italia. Ist. Poligrafico e Zecca dello Stato, Rome.
Arnoldus-Huyzendveld, A., Corazza, A., De Rita, D., and Zarlenga, F. 1997. *Il Paesaggio Geologico ed I Geotopi della Campagna Romana (The Geological Landscape and Geotopes of the "Campagna Romana")*. Fratelli Palombi Editori, Rome. (In Italian and English.)
Chevallier, R. 1976. *Roman Roads.* B. T. Batsford, London.
Funiciello, R., and Thiery, A. 1998. *Il Balcone di Roma.* Fr. Palombi Editori, Rome.
McCall, G. J. H., De Mulder, E. F. J., and Marker, B. R. (eds.). 1996. *Urban Geoscience.* A. A. Balkema, Rotterdam.
Potter, T. W. 1987. *Roman Italy.* University of California Press, Berkeley.
Praturlon, A. 1990. Materiali da Costruzione a Resorse Minerarie. In D. Cosetino, M. Parotto, and A. Praturlon (eds.), 1993. *Guide Geologiche Regionali, 14 Itinerari—Lazio*, pp. 81–88. Società Geologica Italiana, Rome.
Richmond, I. 1930. *The City Wall of Imperial Rome.* Oxford University Press, Oxford.

Chapter 8. The Celian Hill

Ashby, Thomas. 1935. *The Aqueducts of Ancient Rome.* Clarendon Press, Oxford.
Bono, P. 1993. Risorse Idriche. In *Guide Geologiche Regionali, 14 Itinierari—Lazio.* Società Geologica Italiano, Rome.

Bono, P., and Boni, C. 1996. Water Supply of Rome in Antiquity and Today. *Environmental Geology* 27: 126–34.

Cichi, S., Messina, B., and Pocino, W. 1981. *Lazio—Invito Alle Acque.* Istituto Geografico de Agostini, Novara.

Comune di Roma. 1986. *Il Trionfo dell'Aqua—Acque e Acquedotti a Roma—IV sec.a.C-XX sec.* Museo della Civilta Romana and Paleeani Editrice, Rome.

Corrazza, A., and Lombardi, L. 1995. Idrogeologia dell'area del centro storico di Roma. In R. Funiciello (ed.), *La Geologia di Roma—Il Centro Storico,* pp. 179–211. Vol. 50 of Memorie descrittive della Carta Geologica d'Italia. Ist. Poligrafico e Zecca della Stato, Rome.

De Rita, D., Lucchitta, I., and Giampaolo, C. In preparation. Rome, the City That Owes Its Success to Tuffs. Accepted for the *Tuff Handbook* (ed. G. Heiken).

De Rosa, R., Liberati, A.-M., and Pace, P. 1989. *Gli Acquedotti di Roma nell'Epoca Classica.* Provincia di Roma, Rome.

Faccenna, C., Funiciello, R., Montone, P., Parotto, M., and Voltaggio, M. 1994. Late Pleistocene Strike-Slip Tectonics in the Acque Albulae Basin (Tivoli, Latium). *Memorie Descrittive della Carta Geologica díItalia* 49:37–50.

Funiciello, R., and Rosa, C. 1995. L'area romana e lo sviluppo delle richerche geologiche. In Funiciello, R. (ed.), *La Geologica di Roma—Il Centro Storico,* pp. 23–29. Vol. 50 of Memorie descrittive della Carta Geologica d'Italia. Ist. Poligrafico e Zecca della Stato, Rome.

Goethe, J. W. 1962. *Italian Journey [1786–1788].* Translated by W. H. Auden and E. Mayer. Penguin Books, London.

Lombardi, L., and Corazza, A. 1995. *Le Terme di Caracalla.* Fr. Palombi Editori, Rome.

Morton, H. V. 1966. *The Fountains of Rome.* Macmillan, New York.

Chapter 9. The Esquiline Hill

Claridge, A. 1998. *Rome—An Oxford Archaeological Guide.* Oxford University Press, Oxford.

Healy, J. F. 1978. *Mining and Metallurgy in the Greek and Roman World.* Thames and Hudson, London.

Landels, J. G. 1978. *Engineering in the Ancient World.* University of California Press, Berkeley.

Lombardi, L., and Corazza, A. 1995. *Le Terme di Caracalla.* Fr. Palombi Editori, Rome. (If you are visiting the Baths of Caracalla, you should read this book. Even if you do not read Italian, the illustrations are very interesting).

Meiggs, R. 1982. *Trees and Timber in the Ancient World.* Oxford University Press, Oxford.

Panati, C. 1987. *Extraordinary Origins of Everyday Things.* Harper and Row, New York.

Provincia di Roma. 1997. *Relazione Sullo Stato dell'Ambiente a Roma—Anno 1997.* Maggioli Editore, Rome.

CHAPTER 10. THE VIMINAL AND QUIRINAL HILLS

Borghini, G. 1997. *Marmi Antichi.* Edizione DeLuca, Rome. (*Marmi* means "marbles" but is often used in the art world to describe many different rock types used in sculpture and architecture. This remarkable book has examples of many rock types, all used in Rome, but it concentrates mostly on marbles.)

Cavinato, G. P., Funiciello, R., Mattei, M., and Vecchia, P. 1990. Da Passo Corese a Micigliano. In Interario No. 10. In D. Cosentino, M. Parotto, and A. Praturlon (eds.), 1993. *Guide Geologiche Regionali, 14 Itinerari—Lazio,* pp. 246–72. Società Geologica Italiana, Rome.

Funiciello, R., and Mattei, M. 1991. Le Roccia di Faglia nel Barocco Romano. *Le Scienze* 276: 38–45.

Giampaolo, C., De Rita, D., Capicotto, B. M., Godano, R. F., and Di Pace, A. 1998. Un data base sui Litotipi Italiani. *L'Informatore del Marmista,* November, pp. 31–39 (In Italian and English.)

Packer, J. E. 1991. Roman Building Techniques. In M. Grant and R. Kitzinger (eds.), *Civilization of the Ancient Mediterranean,* vol. 1, pp. 299–321. Scribner's, New York.

Passchier, C., and Trouw, R.A.J. 1996. *Microtectonics.* Springer, Berlin.

Rockwell, P. 1993. *The Art of Stoneworking.* Cambridge University Press, Cambridge. (If you have any interest in stoneworking, this is the book to read.)

Yardley, B.W.D. 1989. *An Introduction to Metamorphic Petrology.* Longman Scientific and Technical, New York.

Index

Page numbers in italics refer to illustrative material.

Abruzzo, 102, 137
Acque Albule travertines, 125, 126, *127*
Adriatic Sea, *13*, 14, 22
Aesculapius, 71
Agrippa (emperor), 146, 157, 206
Alban Hills, ix, 7, 8, 15, 35, 44, 47, 51, 65, 71, 79, 121–22, 126, 148, 194, 205, 207; craters of, 33–34, 213, 214–15; history of, 29–34; human occupation of, 30–31; lakes of, 30; size and elevation of, 29–20; vegetation of, 30; vineyards of, 34; volcanic fields of, *10*, 11, 28–29, *29*, *30*, 30, 32–34, 42, 55, 110, *118*, 119, 137, 162, 174–75, *177*, 189, *216*
Albinus, 117
Alexander Severus (emperor), 147
amphorae, 91–92
andosol, 209
Aniene River, 39, 40, 43, 65, 79, 140, 144; and the Aniene River valley, 75, 77
Apaventa, Silvio, 77
Apennines, the, ix, 12, *13*, 14, 19, 26, 105, 110, *139*; central Italian, 138, 140; inhabitants of, 12; rainwater on, 124, 138; as source of earthquake activity, 103, 105–6; types of rock found in, 20, 22, 137; of Umbria, 22, 102; water supply from, 137–38, 140
Apollo and Daphne, 167
Apollodorus of Damascus, 153–54, 192, 200
Appian Way, 116, 119, 210, *212*
Appian, 99
Appius Claudius Caecus, 116, 144
Acqua Paola Fountain, *132*, 176–78, *178*
Aquae Sallustianae, 85
Aqueducts of Ancient Rome (Ashby), 147–48
aqueducts of Rome, 129, 141–43, 176–77, 207; Alexandrina, 147, 148; Anio Vetus, 144, 145; Aqua Alsietina, 146; Aqua Appia, 141–42, 144–45, 146; Aqua Claudia, 103, *145*, 147, 226–28, *227*; Aqua Julia, 130, 146; Aqua Marcia, 145, 148; Aqua Tepula, 146; Aqua Traiana, 147, 148, 176, 222; Aqua Virgo, 2–3, 142, 146, 148, 199; basic components of, *146*; of the Classical period, 143–47; maintenance of, 132–43, 146; Peschiera-Capore, 149–50; restorations of, 143, 148–49; slope of, 143; Vergine, 3. *See also* Baths of Caracalla; springs of Rome
Arch of Dolabella, 126
Arch of Janus, *184*
Arenula Creek, 184
Ashby, Thomas, 147–48
Augustus (emperor), 99
Aurelian (emperor), 113
Aurelian Wall(s), the, 88, 103, 106, 114, *114*; geologic cross section below, *107*
Aventine (Aventino) Hill, 8, 51–52, 58, 184–86, 187, 204; quarry complex underneath, 53

Baccarini, Alfredo, 75, 77
Bacci, Andrea, 75
Bagni di Tivoli, 33, *128*

Basilica of Sant'Agnese in Agone, 171, *172*
Basilica of Saint Peter, 101, 110, 169, 171, 187, 197; construction of on alluvium, 107–8
Basilica of San Giovanni, 101
Basilica of San Paolo fuori le Mura, 65, 100
Basilica of Santa Sabina, 58
basoli, 119, *120*
Baths of Caracalla (Terme di Caracalla), 85, 129–30, 157, 207–8, 224–26; geology of, 226; heating of, 157
Baths of Titus, 54
Baths of Trajan, 54, 94, 153, 189
Bellotti, Piero, 82, 83
Bernini, Giovanni Lorenzo, 3–4, 107, 136, 166–67, 168, 171, 194
Boniface IV (pope), 69
Bono, P., 140
Borghese, Scipione, 167
brick production, 115–16

Caesar. *See* Julius Caesar
Caius Plautius, 144
Calcara, 209
calcium carbonate, 124, 137
Caligula (Emperor), 147, 226
Campanian Plain, 12
Campi di Annibale, 215
Campo de' Fiori, 59, 148
Capitoline (Campidoglio) Hill, 8, 27–28, 36, 71, 85, 192, 193–94, 200, 204; composition of, 194
Capo di Bove lava flow, *118*, 119, 210
Caracalla (emperor), 129
carbon dioxide, 35, 124, 126
carceraes, 206
Cassio, 222
Castel Gandolfo, 28–29
Castel Sant'Angelo, 114, 121
Castellani, Alessandro, 77
Castelli Romani, 28, 215
Castor and Pollux sculptures, 163
Catacombs, the, 55; internment within, 55; of San Callisto, 55–56; of San Sebastiano, 55
Catherine of Siena, Saint, 60
Celian (Celio) Hill, 8, 85, 97, 123, 188–89, 204, 223–24
charcoal, 156–57; importation of, 156
Church of Sant'Anastasia, 224
Church of San Bartolomeo, 67, 71
Church of San Clemente, 223
Church of San Gregorio Magno, 123, 188
Church of San Vitale, 90, *91*
Church of Santa Maria ad Martyres, 69
Church of Santa Maria delle Grazie, 219
Church of Santa Maria in Dominica, 188
Church of Santa Maria Maggiore, 189
Church of Santa Maria Sopra Minerva, facade plaques marking water depths, *61*
Church of Santi Giovanni e Paolo, 123
Church of S. S. Nome di Maria, 94
Cinecittà, 15
Circus Maximus, 29, 37, 51, 85, 91, 97, 158, 185, 186–88, 191, 205–7, *206*; and the "eggs," 206; reconstruction of after fires, 206; seating capacity of, 206
Circus of Gaius and Nero, 106–7, 187
Claudius (emperor), 146, 147, 226
Clement VI (pope), 100
Clement VIII (pope), 60
Clement IX (pope), 219
Cloaca Maxima, 85, 87, 183, 200, 223, 223, 224
Colosseum, the, 37, 47, 95, 97, *98*, 101, 126, 153, 189, *190*, 191; construction of, 97–98; repairs of after earthquake, 96–97
colombaria, 209
Column of Marcus Aurelius, 164, *165*, 197, 198, *198*, 201; alluvial deposits underlying, 166

Column of Trajan. *See* Trajan's Column
concrete, 69, 71
Cornelius Dolabella, 126
Constantine (emperor), 106–7
Constantius II, 187
Conti, Maurizio, 56
Corazza, Angelo, 129–30, 157
Corsica, 22

De aque ductu (Frontinus), 143–44
De Rita, Donatella, xi
Deacon Callisto, 55
Decius Marius Venantius Basilius, 96
Diodorus Siculus, 117
Dionysius of Halicarnassus, 206
dolomite, 137
Domus Augustana, 50
Domus Aurea ("Golden House"), 54, 189

earthquakes, in Rome, 7, 96–99, 100, 108–9, 166, 199; and data for earthquake history, 99; earthquake of 1349, 99–101; earthquake of 1703, 102–3; earthquake of 1915, 103; effects of on groundwater, 102–3; epicenters of in the Italian peninsula, *105*; and the great basilicas of Rome, 104; Mercalli Earthquake Intensity Scale, 101; risk of, 96; sources of, 103, 105–6
Emporium, 91, 169
energy consumption in Rome, 154–56; demand for firewood, 157; electricity consumption, 159; in modern Rome, 158–61; natural energy resources in ancient Rome, 156–58
Engineering in the Ancient World (Landel), 154
Esquiline (Esquilino) Hill, 8, 54, 85, 97, 148, 152, 153–54, 161, 189, 204
Etruscans, the, 13, 40, 44

Faccenna, Claudio, 17, 126
Faete volcano, 118, 119
Farnese, Alessandro, 38
Farnese Gardens, 38
Flaccus, 117
fontanelle, 4
Fortuna Virile, 183
Forum, the. *See* Roman Forum, the
Forum Boarium, 183, 184, 205
Forum of Augustus, 145, 164
Forum of Trajan, 94, 164, 200–1
Fosso dell'Acque Mariana, 184, 186
Fra Angelico, 60
Frascati, 215
Frontinus. *See* Sextus Julius Frontinus
Funiciello, Renato, xi, 97, 98, 106, 169

Gaius Flaminius, 116
Gaius Junius Silanus, 126
Garibaldi, Giuseppe, 114; attempts to divert the Tiber River, 77–79; political cartoon of, *79*
gasometers, 158, 187
Geologia di Roma (National Geological Service of Italy), xi
geology: geological timelines, *21*, *23–25*; introduction to, 2–26. *See also* earthquakes, in Rome; Italy: geology of; Rome: and geological processes
German Academy of Rome, 111
Getty, J. Paul, 170
Giampaolo, Ciriaco, 170
Gianicolo Park, 150
Goethe, Johann Wolfgang von, 28, 130
Gran Raccordo Annulare, 10, 71
Gran Sasso, 12, 22

Hadrian (emperor), 69, 94
Healy, J., 156
Heiken, Grant, xi
Holocene epoch, 33
Honorius, 113

Hospital of Santo Spirito, 219
hydrogen sulfide, 126
hypocaust, 157–58

Imperial Rome, 4, 11, 17, 54, 99, 183, 195; flooding in, 63, 75
Innocent XI (pope), 222
Innocent X (pope), 107, 131
International Centre for the Study of the Preservation and Restoration of Cultured Property, 47
Isola Tiburina (Tiberina Island), 72–74, 78
Italy, 103; geology of, 18–19; oil and natural gas resources of, 159–60; 300-kilometer view of the Italian peninsula, *13*; 300 million to 20 million years ago, 20, 22; 3,600-kilometer view of Italy and the surrounding Mediterranean area, *17*; 20 million years ago to the present, 22, 26

Janiculum (Gianicolo) Hill, 29, 71, 110–11, 148, 175, 178, 197; and alluvial deposits, 106; reservoir on, 150; spring at base of, *180*
Julius II (pope), 107
Julius Caesar (emperor), 75, 83
Julius Obsequens, 35
Juvenal, 99

karst terrains, 138, 140

La Maranella, 65
Lacus Curtius, 85
Lake Albano, 30, 31, *31*, 34, 137; geologic formation of, 214–15
Lake Bolsena, 15
Lake Bracciano, 15, 31, 137, 147, 176
Lake Martignano, 146
Lake Nemi, 30, 31, 34
Lanciani, Rodolfo, 71
Lancisi, 219
Landel, J. G., 154
Lanzani, Don Vittorio, 169
lapis Albanus, 44–45
lapis Gabinus, 44–46
Largo San Rocco, 75
Lazio, 20, 65, 137
Leo IV (pope), 114
Leo X (pope), 60
Ligurian Sea, 22
limestone, 124–25, 137, 149, 209; metamorphosis of to marble, 167–68
Lipari, 88
"Little Ice Age" in Europe, 75, 83
Livy (Titus Livius), 11, 35, 63, 99
Lombardi, Leonardo, 129–30
Lucullo villa, the, 195, 197
Luna (Luni), 169

Marcus Aemilius Lepidus, 144
Marcus Fulvius Nobilior, 144
Marcus Licinius Crassus, 144
Marrana della Caffarella, 65
marble, 16, 169–70; Carrara marble, 4, 22, 170–71; composition of, 167–68; Cottanello marble, 171; marble *breccias*, 168; textures of, 168–69
Marcus Agrippa, 69
Marcus Aurelius, 201
Marino, 40, 213–14
Maxentius, 113
Meiggs, Russell, 157
Michelangelo, 27; copy of his house in Rome, 150
mineral water, composition of, 151, 179
Ministero dell'Interno, 163, 194
Molini, Paolo, 77
Monte Cavo, 215
Monte Giordano, 148
Monte Mario, 110–11
Monte Soratte, 35
Monte Testaccio, 91–94, *92*, *93*
Mount Etna, 28
Mount Vesuvius, 28, 35

Murcia valley (Valle Murcia), 186–87, *186*, 205
Museo delle Mura (Museum of the Walls), 114, 209
Mussolini, Benito, 191, 192

National Hydrographic Service, 63, 67
Nemi, 214
Nero, 54, 97, 189
Nerva (emperor), 144
Nicholas V (pope), 3, 107, 148
Norcia, 102

obsidian tools, 88
olive oil, 92–93
Oppian Hill, 54, 95, 189
Orosius, 35
Ostia Antica, 11, 80–81, *81*

Pace, P., 140
Pagliacci, Federico, 56
Palace of Justice, 94–95
Palatine (Palatino) Hill, 8–9, 36, 37–38, 51, 85, 97, 185, 187–88, *188*, 191, 200, 204; monuments of, 46, 50
Palazzo Borrelli, 94
Palazzo Corsini, 183, 221
Palazzo dei Conservatori, 27, 194
Palazzo del Quirinale, 163, 171, 195
Palazzo Nuovo, 27, 194
Palazzo Senatorio, 27, 45, 194
Palazzo Valentini, 94
Palmarola, 88
Pantheon, the, 8, 59, *70*; and early utilization of concrete, 69, 71
Paolo V Borghese, 148
Parade (Picasso), 177, *179*
Parco di Traiano, 153, 158
Parotto, Maurizio, 19, 20
Parrasio Woods, 178
Passeggiata del Gianicolo, 175
Paul, Saint, 106
Paul II (pope), 148
Paul V (pope), 130, 131, 148, 176
Peter, Saint, 106, 223
Petrach, 100
Piazza Albania, 52, 53
Piazza Barberini, 90, 163
Piazza Cavour, 94–95
Piazza Colonna, 197, 199
Piazza dei Cinquecento, 162–63
Piazza del Campidoglio, 27, 36, 193–94
Piazza del Popolo, 206
Piazza del Porto Ripetta, 61
Piazza Bocca della Verità, 183
Piazza della Navicella, 226
Piazza della Repubblica, 162, 163, 194
Piazza delle Vaschette, 219
Piazza di Porta Capena, 195, 207, 224
Piazza di San Giovanni, 206
Piazza di Spagna, 195, 197
Piazza Navona, 8, *60*
Piazza Risorginento, 219
Piazza Trevi, 3
Piazza Tuscolo, 91
Piazza Venezia, 27, 28, 94
Piazza del Viminale, 163
Piazzale Giuseppe Garibaldi, 110
Piazzale Ostiense, 86, 123
Piazzale Ugo de Malfa, 51
Picasso, Pablo, 177
piezometer, 149
Pincian Hill, 8, 85, 148; geologic cross section of, *135*
Pius VII (pope), 219
Pius IX (pope), 148, 158, 187
Pius XIII (pope), 136
Pleistocene epoch, 106, 121, 209, 219
Pliny the Elder, 35, 83
Pliny the Younger, 11, 83
Pliocene epoch, 106
Plutarch, 83
Po River basin, 159
Pomerium, the, 52
Pons Aemilius, 116
Ponte Cestio, *124*

Ponte Emilio, 180–81. *See also* Pons Aemilius
Ponte Galeria, 121, 175
Ponte Garibaldi, 74
Ponte Palatino, 180
"Ponte Rotto," 116, 180–84, *182*
Port of Claudius, 80, 83
Porta Aurelia, 114
Porta Capena, 226
Porta del Popolo, 103
Porta Flumentana, 63
Porta Furba, 103
Porta Maggiore, 103, 144, 228
Porta Melonia, 91
Porta Metronia, 106
Porta Ostiense, 106
Porta San Pancrazio, 114, 116
Porta San Paolo, 90, 106, 114
Porta San Sebastiano, 114
portents, 35
pozzolan (*pozzolana*), 69, 71
Pozzuoli (Puteoli), 69
Prati di Castello, 75
Praturlon, Antonio, 20
Probus (emperor), 113
Pulvis Puteolanus, 69
Pyramid of Caius Cestius (Piramide), *86*, 90, 114
pyroclastic flow deposits. *See* tuffs

Quintus Marcius Rex, 145
quarries, 1, *41*, 53, 71, *72*, *115*, 210, 213; claystone, 115; Cottanello quarry, *173*; gravel, 120; marble, 169, 170, 171; near the Via dei Laghi, 213–14. *See also* Aventine (Aventino) Hill: quarry complex underneath
quarrying, 126, 201, 213; of older buildings for marble, 170; "room and pillar" technique of, 55
Quirinal (Quirinale) Hill, 5, 7, 8, 85, 163–67, 171–72, 192, 195, 200–201
Quirinal Palace, 5, 7

Republican Rome, 2, 54
Ripetta column, the, 61, *62*, 63, 67
Rockwell, Peter, 5
Roman Forum, the, 37, 44, 46, *203*, 203–4; archeological excavation of, *18*, *89*; monuments in, 46
Roman roads, 12, 14, 116–22; ancient construction details of, 117; construction checklist for, 117; types of blocks used in, 119; use of sand and gravel in, 120–22
Roman sewers, 225. *See also* Cloaca Maxima
Roman walls, 113–16; Aurelian, 88, 103, 106, 113–14; and brick production, 115–16; composition of cores of, 115
Rome, ix, xi–xiii, 1–2; basic geology of the historical center of, *109*; city walls of, 52–53; and debris, 88, 90–95, *90*; decline of commerce in owing to flooding, 80–83; flooding in, 60–61, 63–64, 65, 67, *76* (flood of 1870), 181–83; founding of, 26, 36; 4th-century population estimates for, 141; general hydrologic framework of, *138*; and geological processes, ix, xi, 2, 26, 169; hills comprising the "balcony of Rome," 110–11; and importation of building materials, 16; Opera House in, 194; restoration and conservation of historic monuments in, 108–9; satellite view of 20th-century Rome, *10*; tenements in, 99; terrain of, *x*; threat to from underground cavities, *56*, *57*; threat of volcanic eruptions in, 34–36; volcanoes of, 15, 28–29; and water distribution systems, *150*; water supply and population of, *142*. *See also* earthquakes, in Rome; energy consumption in Rome; Imperial Rome; Republican Rome; Roman roads; Roman sewers; Roman walls; Rome, water supply of; Tiber floodplain, the; Tiber River;

Tiber River basin; tuffs, and Rome; underground Rome
Rome, water supply of, 130, 132–33, 216–17; from the Apennines, 137–38, 140; bicarbonate-alkaline waters, 176–77; and bottled water, 151; future of, 151–52; and "gathering" water, 140; geology of springs, 133–34; mineral, 151, 179; "oligominerale" waters, 176; testing quality of, 140–41; from volcanic springs, 137; water distribution in modern Rome, 149–50. *See also* aqueducts of Rome; springs of Rome
Rovelli, Antonio, 97, 98, 166

Sabatini volcanic fields, 7, 11, 28, *29*, 35, 44, 79, 137, 149, 209
Sabine Mountains, 149
Salvi, Nicola, 1, 4
San Giorgio in Velabro, 205
San Giovanni in Laterano, 186, 187
San Pietro in Montorio, 177
sanpietrini, 4, 119
sand and gravel, 120–21; sources of, 122
Santa Beatrice School, 56
Santa Maria in Trastevere, 178–79, 181, *181*
Santa Maria Maggiore, 161
Santa Maria Sopra Minerva, 59–60, 63
Sant'Andrea al Quirinale, 166, 167, 171, 194
Sant'Orofrio, 219
Sardinia, 22, 88
Septimius Severus (emperor), 50, 205
Servian Wall, the, *43*, 52
"seven hills" of Rome, 7–8, 40, 51, 85, 174–75, *176*, *177*, 204; aerial view of, *9*. *See also individually listed specific hills*
Sextus Julius Frontinus, 143–44, 226
Simbruini Mountains, 140
Sixtus V (pope), 148
Spanish Academy of Rome, 177
Specchi, Alessandro, 61–62
Sperone, 47
springs of Rome, 133–36, *134*; Acqua Angelica, 217, 2192; Acqua Damasiana, 136; Acqua della Fontana degli Api, 136; Acqua di Mercurio, 224, *225*; Acqua di San Clemente, 135, 224; Acque di San Felice, 136; Acqua di Santa Maria delle Grazia, 136, 217, *220*; Acqua Innocenziana, 222; Acqua Lancisiana, 136, 219–21, *221*; Acqua Lautole, 135; Acqua Pia, 136, 217, *218*; Acque Corsiniane, 136, 221–2; Acque Sallustiane, 136, 163; at the base of Vatican Hill, 217; Fonte delle Camene, 135; Lacus Juturnae, 223; Lupercale, 136; Piscina Pubblica, 135; Spring of the Temple of the Syrian Goddess, 222–23; Tullianum, 134, 223. *See also* aqueducts of Rome
stone construction, 213; and volcanic ash, 201
subduction zone, 22

Tabularium, 45
Tarpeian Rock, the, 27, 204–5; geologic composition of, 204
Tarquinius Priscus, 206
Temple of Antoninus and Faustina, 45, *45*, 48, *203*, 213
Temple of Claudius, 123, 188
Temple of Cybele, 38
Temple of the Deified Julius Caesar, 44
Temple of the Deified Trajan, 94
Temple of Fortuna Virilis, 183
Temple of Hercules the Conqueror, 183
Temple of Jupiter, 36
Temple of Mars Ultor, 46, 164
Temple of Portunus, 183
Temple of the Syrian Goddess, 222
Temple of Vesta, 183
Termini Station, 52, 90–91
Theater of Marcellus, 8

thermal baths (*thermae*), 129, 157. *See also* Baths of Caracalla
thrust faults, 12, 14, 22
Tiber floodplain, the, 59–64, 164, 221; and the causes of frequent flooding, 65, 67; and climate change, 75; and the definition of a river delta, 81–83; and flooding in Rome, 61, 63–64, 67; narrowing of, 184
Tiber River, xii, 4, 7, 11, 15–16, 26, 32, 39, 43, 44, 58, 94–95, 166, 175, 197, 219; length of, 65; pollution of, 136–37, 142; projects to divert the river, 75, 77–79, *80*; and riverbed erosion, 183; sand and gravel deposits of, 121; tributaries of, 65, 85, 87, 169–70, 175. *See also* Tiber floodplain, the; Tiber River basin
Tiber River basin, *66*; distribution of rainfall over, *68*
Tiber Island, 71–72, *73*, 74, *74*, 116, 179, 184, *185*
Tiburtini-Cornicolani Mountains, 125
Titus Livius. *See* Livy
Tivoli, 1, 124, 125, 144
Tomb of Caecilia Metalla, 119, 209, 210, *211*, 213
Tomb of Sister Maria Raggi, 168
Tomb of the Scipios (Tomba degli Scipioni), *42*, *54*, 208–9
Tor di Nona, 183
Tor Marancia River, 65
Trajan (emperor), 99, 201, 206
Trajan's Column, 94, 163, 164, *165*, 166, *202*; sandstone underlying, 199, 201
Trajan's Forum. *See* Forum of Trajan
Trajan's Markets, 164, 200, *200*, 201
Trastevere, the, 59, 177, 183
Trastevere neighborhood, 8, 88, 111
travertine, 1, 5, 124–27, 194; formation of, 125; location of, 125–26
Trevi Fountain, *xvi*, 1, *3*, 19, 163, 171, 195; construction of new fountain, 4–5; destruction of original fountain by Bernini, 3–4; neighborhood surrounding, *6*; original fountain, 3; as source of fresh water, 2–4; stone used in, 5
Trinità dei Monti, 195, *196*, 197
tuffs, 7, 8, 28, 32, 37–38, *48*, *49*, 51, 87, 106, 121–22, 185, 186, 191, 192, 194, 197; "Cappellaccio," 201; definition of, 38–39; erosion of, 189; excavation of, 54–55; *Lionato* tuff, 45–46, 204; *peperino* tuff (of Marino), 44, 201, 210, 213–15; preservation of in Roman monuments, 47–50; *Tufo di Monteverde*, 46–47; *Tufo Giallo*, 44; *Tufo pisolitico* (pisolitic tuff or "pisolites"), 42–43, 52, 201; welded tuffs, 39. *See also* tuffs, and Rome
tuffs, and Rome, 39–40, 42; usage of in Rome, 42–47
Tullianum fountain, 223
Tuscolano-Artemisio volcano and caldera, 42, 47, 204–5, 209, 215; and the Faete cone, 210, 214, 215
Tuscany, 65, 88
Tuscolo (Tusculum), 215
Twain, Mark, 28
Tyrrhenian Sea, 4, 14–15, 82

Umbria, 20, 22, 65
underground Rome, 52–58; as a cultural heritage, 57–58; tunnel complexes, 56–57. *See also* Catacombs, the
University Botanical Garden, 178

Valentini, Vincenzo, 94
Valle delle Camene, 85, 224
Valle d'Inferno, 75
Valley of Almone, 77
Varco di San Giovanni, 77
Vatican, the, 106–8, 148, 169, 183, 217. *See also* Basilica of St. Peter; Vatican Hill

Vatican Hill, 110–11; springs at the base of, 217
Velabrum, 183, 184, 205
Velabrum Minus, 85
Velletri, 214
Vespasian (emperor), 97
Vesta, 183
Via Appia, 14, 44
Via A. Saffi, 113
Via Ardeatina, 65
Via Aurelia, 114, 116, 181
Via Aurelia Antica, 147
Via Barberini, 85
Via Cavour, 85, 91, 163, 194
Via Claudia, 126, 188
Via Condotti, 197
Via Dandolo, 111
Via dei Fori Imperiali, 191–92, 200; "scenic walls" along, *192*
Via dei Laghi, 213–14
Via dei Serpenti, 163
Via del Corso, 197
Via del Monte Caprino, 27
Via del Tarpeo, 36
Via del Teatro di Marcello, 27
Via del Tempio di Giove, 27, 204
Via del Tritone, 7, 85
Via del Velabro, 205
Via della Consolazione, 36, 192–93; tuff outcrops along, *193*
Via dell'Arco Travertino, 56
Via della Lungaretta, 181
Via delle Sette Chiese, 56
Via di Fioranello, 210
Via di Porta San Pancrazio, 178
Via di San Gregorio, 37, 85, 123, 191
Via di San Paolo della Croce, 126
Via di San Sebastiano, 209
Via Flaminia, 14, 197
Via Garibaldi, 178
Via Jugaro, 27
Via Labicana, 54, 85, *120*
Via Marmorata, 170
Via Nazionale, 163, 194
Via Nova, 226
Via Ostiense, 90
Via Prenestina, 44, 144
Via Tiburtina, 46
Via Vittorio Veneto, 85
Viale dei Quattro Venti, 111
Viale del Parco del Celio, 188
Viale del Monte Oppio, 153
Viale delle Terme di Caracalla, 207, 225
Viale di Trastevere, 74
Viale Gianicolo, 220–21
Villa Borghese Park, 8
Villa Celimontana, 126
Villa of Lucullus, 195–96
Villa Sciarra, 111, *112*, 113; geologic cross section through, *113*
Viminal (Viminale) Hill, 8, 85, 88, 162–63, 194
Vitruvius, 71, 129, 140, 141
Vittorio Emanuele Monument, 194
volcanism, 15, 26, 125, 126; hydrovolcanic eruptions, 39; and pyroclastic flows, 32–33, 37

Wright, Joseph, 28